Visualization in the Age of Computerization

Digitalization and computerization are now pervasive in science. This has deep consequences for our understanding of scientific knowledge and of the scientific process, and challenges longstanding assumptions and traditional frameworks of thinking of scientific knowledge. Digital media and computational processes challenge our conception of the way in which perception and cognition work in science, of the objectivity of science, and the nature of scientific objects. They bring about new relationships between science, art and other visual media, and new ways of practicing science and organizing scientific work, especially as new visual media are being adopted by science studies scholars in their own practice. This volume reflects on how scientists use images in the computerization age, and how digital technologies are affecting the study of science.

Annamaria Carusi is Associate Professor in Philosophy of Medical Science and Technology at the University of Copenhagen.

Aud Sissel Hoel is Associate Professor in Visual Communication at the Norwegian University of Science and Technology.

Timothy Webmoor is Assistant Professor adjunct in the Department of Anthropology, University of Colorado, Boulder.

Steve Woolgar is Chair of Marketing and Head of Science and Technology Studies at Said Business School, University of Oxford.

Routledge Studies in Science, Technology and Society

Visualization in the Age of Computerization

**Edited by
Annamaria Carusi, Aud Sissel Hoel,
Timothy Webmoor and Steve Woolgar**

Routledge
Taylor & Francis Group

LONDON AND NEW YORK

First published 2015
by Routledge

2 Park Square, Milton Park, Abingdon, Oxfordshire OX14 4RN
52 Vanderbilt Avenue, New York, NY 10017

*Routledge is an imprint of the Taylor & Francis Group,
an informa business*

First issued in paperback 2020

Library of Congress Cataloging-in-Publication Data
Visualization in the age of computerization / edited by Annamaria Carusi,
Aud Sissel Hoel, Timothy Webmoor and Steve Woolgar.
      pages cm. — (Routledge studies in science, technology and society
; 26)
    Includes bibliographical references and index.
  1.   Information visualization. 2.   Digital images. 3.   Visualization.
  4.   Computers and civilization.   I. Carusi, Annamaria.
    QA76.9.I52V574 2014
    303.48'34—dc23
    2013051326

ISBN13: 978-0-415-81445-4 (hbk)
ISBN13: 978-0-367-60041-9 (pbk)

Typeset in Sabon
by IBT Global.

Contents

PART II
Doing Visual Work in Science Studies

Figures

Introduction

Annamaria Carusi, Aud Sissel Hoel,
Timothy Webmoor and Steve Woolgar

The pervasive computerization of imaging and visualizing challenges us to question what changes accompany computerized imagery and whether, for instance, it is poised to transform science and society as thoroughly as the printing press and engraving techniques changed image reproduction (Eisenstein 1980; Rudwick 1976), or as radically as photography altered the conditions of human sense perception and the aura of works of art (Benjamin [1936] 1999). Whereas some scholars discern continuity in representational form from Renaissance single-point perspective to cinematic and digital arrays, from Alberti's windows to Microsoft (Crary 1990; Manovich 2001; Friedberg 2006; Daston and Galison 2007), other scholars understand our immersion in imagery as heralding a "visual turn" with the engagement of knowledge in contemporary culture (Rheingold 1992; Mitchell 1994; Stafford 1996). James Elkins suggests that visual literacy spans the specialized disciplines of the academy (Elkins 2007), while Robert Horn or Thomas West claim that "visual thinking" is primary and intuitive (Horn 1998; West 2004). Computational high-tech may be enabling the reclamation of old visual talents devalued by word-bound modernist thought.

When we organized the Visualization in the Age of Computerization conference in Oxford in 2011, we were motivated by curiosity regarding what specific effects these innovations were having and whether claims regarding new and more effective visualizing techniques and transformed modes of visual thinking were being borne out. A further motivation was the conviction that the exploration of the field would be enriched by input from the broad range of science studies, including sociological, philosophical, historical and visual studies approaches. The papers presented at the conference and the further discussion and debate that they engendered have gone some way in providing a set of crossdisciplinary and sometimes interdisciplinary perspectives on the ways in which the field of visualization has continued to unfold. This volume gathers together a sample of contributions from the conference, with others being presented in the special issue of *Interdisciplinary Science Reviews*, titled "Computational Picturing."[1] Other interventions at the conference have been published elsewhere, and since that time, the domain has garnered the interest of a great many scholars across the disciplines. New publications in the area include Bartscherer and Coover (2011), Yaneva (2012), Wouters et al. (2013) and Coopmans et al. (2014). The texts gathered in this volume each address the ways in which the increasing prevalence of computational means of imaging and visualizing is having an impact on practices and understandings of "making visible," as well as of the objects visualized. We note five overlapping themes in considerations of the ways in which the modes of visualization associated with digital methods are making a difference in areas that have traditionally been of interest to studies of the visual in science.

First, a difference is being made to the deployment of perception and cognition in visual evidence. Traditional ideas about cognition have long been rejected in favor of an understanding of interpretation in terms of *in situ* material practices. It is thus recognized that mentalistic precepts such as "recognizing patterns," "identifying relationships," "assessing fit and correspondence," etc. are better treated as idealized depictions of the activities involved in generating, managing and dealing with representations. In science and technology studies (STS), visual studies, history of science and related disciplines, the focus on material practices resulted in a widespread acknowledgment of the muddled and contingent efforts involved in the practical activities of making sense of the visual culture of science (Lynch and Woolgar 1990; Cartwright 1995; Galison 1997). Yet the advent of new modes of computerized visualization has seen the reemergence of a mentalistic cognitivism, with visualizations assigned the role of "cognitive aids" (Bederson and Schneiderman 2003; Card, Mackinley and Schneiderman 1999; McCormick, DeFanti and Brown 1987; Ware 2000; see Arias-Hernandez, Green and Fisher 2012 for a review). However, alongside the reemerging cognitivist discourse, there are other claims that accentuate the cognitive import of visualizations— that is, their active role in forming and shaping cognition—while simultaneously pointing to the complex embedding of perception and cognition in material settings (Ellenbogen 2008; Carusi 2008; Hoel 2012). As Anne Beaulieu points out in her contribution toward the end of this volume, even if the term "visual" relies on the perceptual category of sight, vision is networked and witnessed by way of computational tools that take primarily nonoptical forms. Lisa Cartwright too calls attention to the "always already" material aspects of the visible, pointing out the extent to which a materialist philosophy has been developed in feminist approaches to images, imaging and visualization. The cognitive import of visualizations is emphasized in the chapter by Matt Edgeworth in his discussion of discovery through screen-work in archaeology, and in the chapter by Tom Schilling in the context of antagonistic communities using maps or data sets as visual arguments. Alma Steingart, likewise, underscores the cognitive import of visualizations by tracking the way that computer graphics animations were put to use by mathematicians as generative of novel theories. In fact, most contributions to this volume lay stress upon the cognitive import of visualizations, by pointing to their transformative and performative roles in vision and knowledge.

A second change is in the understanding of objectivity in the methodological sense of what counts as an objective scientific claim. Shifts in observational practice are related to fluctuations as to what counts as "good" vision with respect to the type of vision that is believed to deliver objectivity. However, there are diverse notions of objectivity in the domain mediated by computational visualization. Scholars have described the involvement of technologies in science in terms of the historical flux of epistemic virtues in science, tracing a trajectory from idealized images that required intervention and specialized craftsmanship to more mechanical forms of recording. A later hybridization of these virtues has been claimed to occur when instruments and, later, computers allowed active manipulation versus passive observation (Hacking 1983; Galison 1997). Other researchers suggest that the computerization of images accounts for an epistemic devaluation of visualizations in favor of their mathematical manipulation that the digital (binary) format allows (Beaulieu 2001). This raises questions of objectivity in science, but not without also raising questions about the objects that are perceived through different modes of visualization (Hoel 2011; Carusi 2012). In this volume, Chiara Ambrosio shows that notions of scientific objectivity have been shaped in a constant dialogue with artistic representation. She points to the importance of contextualizing prevalent ideas of truth in science by attending to the roles of artists in shaping these ideas as well as challenging them. Annamaria Carusi and Aud Sissel Hoel develop an approach to visualizations in neuroscience that accentuates the generative dimension of vision, developing Maurice Merleau-Ponty's notion of "system of equivalences" as an alternative to approaches in terms of subjectivity or objectivity. The generative roles of visualizations, or, more precisely, of digital mapping practices, are also emphasized in Albena Yaneva's contribution, as well as in the contribution by Torben Elgaard Jensen, Anders Kristian Munk, Anders Koed Madsen and Andreas Birkbak.

A third area where we are seeing shifts is in the ontology of what counts as a scientific object. The notion of what constitutes a scientific object has received a great deal of attention for quite some time in science studies broadly (philosophy, history and sociology). However, we are currently seeing a resurgence of interest in the ontology of scientific objects (Barad 2007; Bennett 2010; Brown 2003; Harman 2009; Latour 2005; Olsen et al. 2012; Woolgar and Lezaun 2013; among many others; see Trentmann 2009 for an overview). As Cartwright points out in her chapter, a new preoccupation with ontology in science studies has been continuous with the digitalization of medicine, science and publication. It may seem that this is because computational objects demand a different sort of ontology: They are often seen as virtual as opposed to physical objects, and their features are often seen as in some sense secondary to the features of "actual" physical objects that they are assumed to mimic. However, the distinction between virtual and physical objects is problematic (Rogers 2009; Carusi 2011; Hoel and van der Tuin 2013), and is inadequate to the working objects handled and investigated by the sciences. This is highlighted when attention is paid to the mode of visualization implied by digitalization and computerization. In this volume, Timothy Webmoor observes that the formerly distinct steps of crafting visualizations are woven together through the relational ontology of "codework." A relational ontology is also key to the approach developed by Carusi and Hoel, as it is to other contributions that emphasize the generative aspect of visualizations. Steingart, for example, points out that the events portrayed in the mathematical films she has studied are not indexes of observable natural events, but events that become phenomena at all only to the extent that they are graphically represented and animated. However, the ontological implications of visualization are frequently obscured: Michael Lynch and Kathryn de Ridder-Vignone, in their study of nanotechnology images, note that by resorting to visual analogies and artistic conventions for naturalistic portrayal of objects and scenery, nano-images represent a missed opportunity to challenge conventional, naturalistic ontology.

In terms of the ontology of visual matter, some visual culturalists and anthropologists (e.g., Gell 1998; Pinney 1990, 2004) have suggested that taking ontology seriously entails reorienting our epistemic toolbox for explaining images, or suspending our desire for epistemic explanation. These include "scaling-out" strategies that move away from the visual object to emphasize historical contingency, partiality or construction through documenting an object's many sociotechnical relationships. A different mode of engagement may be localizing at the scale of the visual object to get at the "otherness," "fetishization" and inherent qualities that make certain visual imagery work (Harman 2011; Henare, Holbraad and Wastell 2007; Pinney 1990; Webmoor 2013; topics also discussed by Beaulieu in this volume). Deploying media to register ontological competencies, rather than to serve epistemic goals, means developing object-oriented metrologies, or new methods of "measuring" materiality (Webmoor 2014). This tactic would in part strive for the affective appreciation of visual objects, and it may approximate the collaboratively produced imagery of artists and scientists. Hence, we view the increase in collaborations between science and art, or alternatively, between science/visual studies and art, as a fourth area of change. Certainly, as has been well documented, there is a long history of productive interchange between scientists and artists (Stafford 1991; Jones and Galison 1998; Kemp 2006). Further, the introduction of digital and computational technologies has opened a burgeoning field of explorations into embodied experience and technology, challenging boundaries between technoscience, activism and contemporary art (Jones 2006; da Costa and Philip 2008). In fact, one of the things that we were aiming for with the conference was to encourage interdisciplinary dialogue and debate between different approaches to science and image studies, and particularly to get more conversations and collaborations going between STS and the humanities, as well as between science/visual studies and creative practitioners. To attain the second goal, artists and designers, including Alison Munro, Gordana Novakovic and Alan Blackwell, were invited to present their work at the conference. In this volume, relations between science and art are dealt with by several contributors. Ambrosio discusses three historical and contemporary examples, where science has crossed paths with art, arguing that an adequate understanding of the shifting notions of objectivity requires taking into account the variegated influences of art. Carusi and Hoel show the extent to which a scientific practice such as neuroimaging shares characteristics of the process of painting, while David Ribes shows how today's visualization researchers apply insights from visual perception and artistic techniques to design in order to ensure a more efficient visualization of data. The chapter by Lynch and de Ridder-Vignone also testifies to the interplay of science and art, not only through the reliance of nano-images on established artistic conventions, but also due to the way that nano-images are often produced and displayed as art, circulated through online "galleries."

A fifth change relates to the way that computational tools bring about modifications to practice, both in settings where visualizations are developed and used in research, and in settings where scholars study these visualizing practices. A series of studies have documented the manner in which visually inscribing phenomena makes up "the everyday" of science work (Latour and Woolgar 1979; Lynch and Woolgar 1990; Cartwright 1995; Galison 1997; Knuuttila, Merz and Mattila 2006), suggesting that much of what enables science to function is making things visible through the assembling of inscriptions for strategic purposes. However, whereas previously researchers investigating these visual inscription processes made their observations in "wet-labs" or physical sites of production, today they often confront visualizations assembled entirely in the computer or in web-based "spaces" (Merz 2006; Neuhaus and Webmoor 2012). Several contributors explore how tasks with computational tools of visualization are reconfiguring organizational and management practices in research settings. Ribes, focusing on visualization researchers who both publish research findings and develop visualization software, shows how, through visual input, technology takes on a leading role in sustaining collaborations across disciplines, which no longer depend solely on human-to-human links. Webmoor ethnographically documents changes to the political economy of academia that are attendant with the cocreation of data, code and visual outputs by researchers, software programmers and lab directors. He poses this reflexive question: What changes for academics in terms of promotion, assignation of publishing credit and determination of "impact factors" will accompany the increasing online publishing of research outputs, or visualizations, in digital format? Modifications to practice due to new computational tools is also a key topic in Edgeworth's chapter, which tracks the transition from spade-work to screen-work in archaeology, as it is in Schilling's chapter, which points to the blurring of boundaries between producers and users in today's mapmaking practices.

When it comes to the modifications to the second setting, to the way that today's scholars go about investigating the practices of science, Peter Galison's contribution to this volume challenges scholars to *do* visual work and not only to study it. Galison's chapter, however, is of a different kind than the preceding chapters, being set up as a debate contribution advocating film as a research tool in science and technology studies—or what he refers to as "second-order visual STS." Upon receiving this contribution, we had the idea of inviting scholars of the visual culture of science to respond, or alternatively, to write up their own position pieces on the topic of visual STS. Thus, whereas the first part of the book consists of a series of (in Galison's terms) first-order VSTS contributions, the second part consists of a series of debate contributions (varying in length from short pieces to full length papers) that discuss the idea of visual STS, and with that, the idea of *doing* visual work in science studies (second-order VSTS). This was particularly appropriate since, as the book was being finalized, we noted that the call for submissions for the 2013 Annual Meeting of the Society for Social Studies of Science included a request for "sessions and papers using 'new media' or other forms of new presentation," and that there would be "special sessions on movies and videos where the main item submitted will be a movie or video."[2] This initiative has been followed up by the announcement of a new annual film festival, Ethnografilm, the first of which to be held in Paris, April 2014. The expressed purpose of the film festival is to promote filmmaking as a "legitimate enterprise for academic credit on a par with peer reviewed articles and books."[3] Clearly, science/visual studies scholars are themselves experiencing the challenge of digital technologies to their own research and presentation practice, and are more than willing to experiment with using the visual as well as studying it. As many of the contributors to the second part of the book discuss, conducting research through filmmaking or other visual means is not unprecedented—Galison himself discusses visual anthropology, and Cartwright discusses the use of film by researchers such as Christina Lammer. However, digital tools make movies and videos available to a far wider range of scholars (consider Galison's discussion of the extremely time-consuming editing process in predigital filming), and also make other techniques that have been used in the past—such as mapping and tracing networks—more available. This is not only a question of availability: Digitalization also transforms these techniques and media, and the mode of knowledge attained through them, as is clearly seen in the discussion of mapping in the responses of Jensen et al. and Yaneva. Besides, the increasing availability of digital video challenges science studies scholars not to fall back into what Cartwright terms an "anachronistic realism", but to deploy new modes of reflexivity in their use of the medium.

With the call for second-order VSTS, and the merging of crafts formerly divided between the arts and sciences, we note an optimistic potential for the engagement with visualizations in the age of computerization. With computerized forms of capture, rendering, distribution and retrieval of scholarly information, the repertoires of science as well as science/visual studies expand well beyond the historical trajectory of what paper-based media permit. As several contributors urge, visual media with their unique affordances supplement paper-based media and allow complementary and richer portrayals of the practices of science. Further, with this expansion come new forms of knowledge. Certainly, this volume cannot exhaust a field already acknowledged for its inventiveness of new tools and techniques, and is but an initial exploration of some of its central challenges that will no doubt continue to elicit the attention of science studies scholars. What we hope to have achieved by gathering together these studies of computerized visualization is that the *visual* dimension of visualization warrants attention in its own right, and not only as an appendage of digitalization.

OVERVIEW

The present volume is divided into two parts. The first part consists of eight chapters that examine the transformative roles of visualizations across disciplines and research domains. All of these chapters are based on presentations made at the 2011 conference in Oxford. The second part considers instead the use of the visual as a medium in science/visual studies, and consists of two full-length chapters and three short position pieces. The contributors to this part were also affiliated with the Oxford conference in one way or another: as keynote, invited keynote, respondent or participants. The following paragraphs provide a brief overview of all contributions.

Timothy Webmoor, in his chapter "Algorithmic Alchemy, or the Work of Code in the Age of Computerized Visualization," offers an ethnography of an academic-cum commercial visualization lab in London. Work with and reliance upon code is integral to computerized visualization. Yet work with code is like Nigel Thrift's "technological unconscious" (Thrift 2004). Until quite recently it has remained in the black boxes of our computerized devices: integral to our many mundane and scientific pursuits, yet little understood. Close ethnographic description of how code works suggests some novelties with respect to the tradition of examining representation in science and technology studies. Webmoor argues that formerly separated or often sequential tasks, principally data sourcing, programming and visualizing, are now woven together in what researchers do. This means that previously separated roles, such as those of researcher and programmer, increasingly converge in what he terms "codework," an activity resembling reiterative knot-making. Outputs or visualizations, particularly with the "mashed-up" web-based visualizations he studies, are held only provisionally like a knot before they are redone, adjusted, updated or taken down. Nevertheless, codework demonstrates vitality precisely because it confounds conventional schemes of accountability and governance. It plays on a tension between creativity and containment.

Matt Edgeworth's "From Spade-Work to Screen-Work: New Forms of Archaeological Discovery in Digital Space" undertakes an ethnography of practices in archaeology involving the integration of digital tools. A presentation of his own intellectual development in terms of deploying digital tools as an archaeologist parallels and grants a candor to his empirical observations surrounding changes to a field that is often caricatured as rather a-technological due to the down-and-dirty conditions of archaeological fieldwork. Among other new tools of archaeology, Edgeworth underscores the embodiment and tacit skill involved with imaging technologies, particularly those of Google Earth and LiDAR satellite images. He makes the case that screen-work, identifying and interpreting archaeological features on the computer screen, entails discovery in the quintessential archaeological sense, that of excavating pertinent details from the "mess" or inundation of visual information. He proceeds to ask whether these modes of discovery are drastically different, or whether the shift to digital techniques is a step-wise progression in the adoption of new tools. In moving the locus of archaeological discovery toward the archaeological office space, Edgeworth brings up a fundamental issue of identity for "the discipline of the spade" in the digital age.

In his chapter "British Columbia Mapped: Geology, Indigeneity and Land in the Age of Digital Cartography," Tom Schilling offers a detailed consideration of the practices, processes and implications of digital mapping by exploration geologists and Aboriginal First Nations in British Columbia, Canada. In both cases the communities produce maps that enter the public domain with explicit imperatives reflecting their economic and political interests. Exploration geologists use digital mapping tools to shape economic development, furthering their search for new mineral prospects in the Canadian Rocky Mountains. First Nations, on their side, produce digital maps to amplify claims to political sovereignty, by developing data layers consisting of ethnographic data. The chapter also explores the specificities of digital cartography compared to its paper-based predecessors. While both digital and paper-based cartography are invariably political, digital mapping tools are distinguished by their manipulability: By inviting improvisational reconstruction they challenge the distinction between producers and users of maps. However, as Schilling's case studies make clear, these practices have yet to fulfill the ideals of collaborative sharing of data and democratic participation. Further, as community-assembled data sets are taken up by others, the meaning and origins of databases become increasingly obscured.

The contribution by David Ribes, "Redistributing Representational Work: Tracing a Material Multidisciplinary Link," turns our attention to the way in which scientists' reliance on visualization software results in a redistribution of labor in terms of knowledge production, with computer scientists playing a key intermediary role in the process. The chapter follows the productive work of one computer scientist as she built visualization tools for the sciences and medicine using techniques from experimental psychology. He shows that the methods and findings of the computer scientists have two trajectories: On one hand, they are documented in scholarly publications, where their strengths and weaknesses are discussed; on the other, research outcomes also independently inform the production of future visualization tools, and become incorporated into a process of scientific knowledge production in another field. Ribes explores the gap between these two trajectories, showing that as visualization software comes to be used and reused in different contexts, multidisciplinarity is loosened from human-to-human links, and instead becomes embedded in the technology itself.

Michael Lynch and Kathryn de Ridder-Vignone, in their chapter "Making the Strange Familiar: Nanotechnology Images and Their Imagined Futures," examine different types of images of nanoscale phenomena. Images play a prominent role in the multidisciplinary field of nanoscience and nanotechnology; and to an even greater extent than in other research areas images are closely bound to the promotion and public interface of the field. Examining nano-images circulated through online image galleries, press releases and other public forums, Lynch and de Ridder-Vignone make the observation that, even if they portray techno-scientific futures that challenge the viewer's imagination, nano-images resort to classic artistic conventions and naturalistic portrayals in order to make nanoscale phenomena sensible and intelligible—pursuing the strategy of "making the strange familiar." This may seem ironic, since the measurements of scanning-tunneling microscopes are not inherently visual, and the nanoscale is well below the minimum wavelengths of visible light. Nonetheless, the conventions used take different forms in different contexts, and the chapter proceeds to undertake an inventory of nano-images that brings out their distinct modes of presentation and their distinct combinations of imagination and realism.

With the chapter by Chiara Ambrosio, "Objectivity and Representative Practices across Artistic and Scientific Visualization," we turn to the history of objectivity in art and science, and specifically to the way in which they are interrelated. She shows that scientific objectivity has constantly crossed paths with the history of artistic representation, from which it has received some powerful challenges. Her aim is twofold: firstly to show the way in which artists have crucially contributed to shaping the history of objectivity; and secondly to challenge philosophical accounts of representation that not only are ahistorical but also narrowly focus on science decontextualized from its conversations with art. Ambrosio's discussion of three case studies from eighteenth-century science illustration, nineteenth-century photography and twenty-first-century data visualization highlight the importance of placing current computational tools and technologies in a historical context, which encompasses art and science. She proposes a historically grounded and pragmatic view of "representative practices," to account for the key boundary areas in which art and science have complemented each other, and will continue to do so in the age of computerization.

Annamaria Carusi and Aud Sissel Hoel, in their chapter "Brains, Windows and Coordinate Systems," develop an account of neuroimaging that conceives brain imaging methods as at once formative and revealing of neurophenomena. Starting with a critical discussion of two metaphors that are often evoked in the context of neuroimaging, the "window" and the "view from nowhere," they propose an approach that goes beyond contrasts between transparency and opacity, or between complete and partial perspectives. Focusing on the way brain images and visualizations are used to convey the spatiality of the brain, neuroimaging is brought into juxtaposition with painting, which has a long history of grappling with space. Drawing on Merleau-Ponty's discussion of painting in "Eye and Mind," where he sets forth an integrated account of vision, images, objects and space, Carusi and Hoel argue that the handling and understanding of space in neuroimaging involve the establishment of a "system of equivalences" in the terms of Merleau-Ponty. Accentuating the generative dimension of images and visualizations, the notion of seeing according to a system of equivalences offers a conceptual and analytic tool that opens a new line of inquiry into scientific vision.

In her chapter "A Four-Dimensional Cinema: Computer Graphics, Higher Dimensions and the Geometrical Imagination," Alma Steingart investigates the way that computer graphic animation has reconfigured mathematicians' visual culture and transformed mathematical practice by providing a new way of engaging with mathematical objects and theories. Tracking one of the earliest cases in which computer graphics technology was applied to mathematical work, the films depicting four-dimensional surfaces by Thomas Banchoff and Charles Strauss, Steingart argues that computer graphics did more than simply represent mathematical phenomena. By transforming mathematical objects previously known only through formulas and abstract argument into perceptible events accessible to direct visual investigation, computer graphic animation became a new way of producing mathematical knowledge. Computer graphics became a tool for posing new problems and exploring new solutions, portraying events that would not be accessible except through their graphic representation and animation. It became an observational tool that allowed mathematicians to see higher dimensions, and hence a tool for cultivating and training their geometrical imagination.

The focus on film as a mode of discovery makes a nice transition to the next chapter, by Peter Galison, which introduces the second part of the book. Galison challenges science studies to use the visual as well as to study it, and delineates the contours of an emerging visual science and technology studies or VSTS in a contribution with two parts: a theoretical reflection on VSTS, and a description of his own experience doing science studies through the medium of film, in such projects as *Ultimate*

Weapon: The H-Bomb Dilemma (Hogan 2000) and *Secrecy* (Galison and Moss 2008). Galison draws a distinction between first-order VSTS, which continues to study the uses that scientists make of the visual, and second-order VSTS, which uses the visual as the medium in which it conducts and conveys its own research. He proposes that exploring the potential for second-order VSTS is a logical further development of what he holds up as the key accomplishment of science studies in the last thirty years: developing localization as a counter to global claims about universal norms and transhistorical markers of demarcation.

In their collective piece, "Expanding the Visual Registers of STS," Torben Elgaard Jensen, Anders Kristian Munk, Anders Koed Madsen and Andreas Birkbak respond to Galison's call to expand the visual research repertoire by advising an even larger expansion: Why stop at filmmaking when there are also other visual practices that could be taken up by second-order VSTS? Focusing in particular on the practice of making digital maps, they argue that the take-up of maps by second-order VSTS would have to be accompanied by a conceptual rethinking of maps, including a discussion of the way maps structure argumentation. They also pick up on Galison's observations concerning the affectivity of film and video by suggesting that maps also leave the viewer affected through his or her own unique mode of engagement.

In her response, "Mapping Networks: Learning from the Epistemology of the 'Natives,'" Albena Yaneva starts out by pointing to the role of ethnographic images in shaping the fieldworker's explanations and arguments. Further, with the introduction of digital mapping tools into the STS tool box with large projects such as the controversy mapping collaborative project MACOSPOL, a shift to second-order VSTS has already taken place. Emphasizing the performative force of mapping, Yaneva argues that mapping is not a way of illustrating but a way of generating and deploying knowledge. In her own fieldwork, which followed the everyday visual work of artists, technicians, curators, architects and urban planners, Yaneva experimented with swapping tools with the "natives," learning a lot from their indigenous visual techniques. Drawing on this "organic" development of second-order visual methods gained through ethnographic intimacy, she suggests borrowing epistemologies and methods from the "natives."

In her piece "If Visual STS Is the Answer, What Is the Question?" Anne Beaulieu outlines her position on the topic of visual STS. She starts out by countering claims that STS has always been visual or that attending to the visual is equal to fetishizing it. Defending the continued relevance of talking about a "visual turn" in STS, Beaulieu lists five reasons why the question of VSTS becomes interesting if one attends to its specifics: Images are an emergent entity that remains in flux; visual practices are just as boring as other material practices studied by STS; observation takes many forms and must be learned; vision and images are to an increasing extent networked; and attending to the visual can serve to expand the toolkit of

STS by drawing on resources from disciplines such as media studies, film studies, art history and feminist critiques of visuality.

The concluding contribution to this volume, Lisa Cartwright's "Visual Science Studies: Always Already Materialist," demonstrates the importance of approaching the visual in an interdisciplinary manner. Cartwright points out the attention paid to the materiality of the visual that was a hallmark of Marxist materialist and feminist visual studies from the late 1970s onwards. The chapter highlights the specific contributions of materialist and feminist visual studies in addressing subjectivity and embodiment. The focus of visual studies on materiality does not result in a disavowal of or skepticism toward the visual. Cartwright reminds us that the visual turn of science studies since the 1990s coincided with the digital turn in science and technology, including the use of digital media in the fieldwork of science studies scholars; here the author calls for a more reflexive use of the medium. The chapter concludes with a detailed discussion of two films by the sociologist and videographer Christina Lammer—*Hand Movie 1* and *2*—showing how they present a third way, which neither reduces things to their meanings in the visual, nor reduces images to the things and processes around them; instead, they bring out a more productive way of handling visuality and materiality.

NOTES

1. Volume 37, Number 1, March 2012. All the contributions published in the special issue are freely available for download at the following link: http://www.ingentaconnect.com/content/maney/isr/2012/00000037/00000001. Last accessed 4th November, 2013.
2. http://www.4sonline.org/meeting. Last accessed 4th November, 2013.
3. For more information about the film festival, see: http://ethnografilm.org. Last accessed 4th November, 2013.

REFERENCES

Arias-Hernandez, Richard, Tera M. Green and Brian Fisher. 2012. "From Cognitive Amplifiers to Cognitive Prostheses: Understandings of the Material Basis of Cognition in Visual Analytics." In "Computational Picturing," edited by Annamaria Carusi, Aud Sissel Hoel and Timothy Webmoor. Special issue. *Interdisciplinary Science Reviews* 37 (1): 4–18.

Barad, Karen. 2007. *Meeting the Universe Halfway: Quantum Physics and the Entanglement of Matter*. Durham, NC: Duke University Press.

Bartscherer, Thomas, and Roderick Coover, eds. 2011. *Switching Codes: Thinking through Digital Technology in the Humanities and Arts*. Chicago: University of Chicago Press.

Beaulieu, Anne. 2001. "Voxels in the Brain: Neuroscience, Informatics and Changing Notions of Objectivity." *Social Studies of Science* 31 (5): 635–680.

Bederson, Benjamin B., and Ben Shneiderman. 2003. *The Craft of Information Visualization: Readings and Reflections*. Amsterdam: Morgan Kaufmann.

Bennett, Jane. 2010. *Vibrant Matter: A Political Ecology of Things*. Durham, NC: Duke University Press.

Benjamin, Walter. (1936) 1999. "The Work of Art in the Age of Mechanical Reproduction." In *Visual Culture: The Reader*, edited by Jessica Evans and Stuart Hall, 61–71. London: SAGE.

Brown, Bill. 2003. *A Sense of Things: The Object Matter of American Literature.* Chicago: University of Chicago Press.

Card, Stuart C., Jock Mackinlay and Ben Shneiderman. 1999. *Readings in Information Visualization: Using Vision to Think.* San Francisco: Morgan Kaufmann.

Cartwright, Lisa. 1995. *Screening the Body: Tracing Medicine's Visual Culture.* Minneapolis: University of Minnesota Press.

Carusi, Annamaria. 2008. "Scientific Visualisations and Aesthetic Grounds for Trust." *Ethics and Information Technology* 10: 243–254.

Carusi, Annamaria. 2011. "Trust in the Virtual/Physical Interworld." In *Trust and Virtual Worlds: Contemporary Perspectives*, edited by Charles Ess and May Thorseth, 103–119. New York: Peter Lang.

Carusi, Annamaria. 2012. "Making the Visual Visible in Philosophy of Science." *Spontaneous Generations* 6 (1): 106–114.

Carusi, Annamaria, Aud Sissel Hoel and Timothy Webmoor, eds. 2012. "Computational Picturing." Special issue. *Interdisciplinary Science Reviews* 37 (1).

Coopmans, Catelijne, Janet Vertesi, Michael Lynch and Steve Woolgar. 2014. *Representation in Scientific Practice Revisited.* Cambridge, MA: MIT Press.

Crary, Jonathan. 1990. *Techniques of the Observer: On Vision and Modernity in the Nineteenth Century.* Cambridge, MA: MIT Press.

da Costa, Beatriz, and Kavita Philip, eds. 2008. *Tactical Biopolitics: Art, Activism, and Technoscience.* Cambridge, MA: MIT Press.

Daston, Lorraine, and Peter Galison. 2007. *Objectivity.* New York: Zone Books.

Eisenstein, Elizabeth. 1980. *The Printing Press as an Agent of Change: Communications and Cultural Transformations in Early Modern Europe.* Cambridge: Cambridge University Press.

Elkins, James. 2007. *Visual Practices across the University.* Munich: Wilhelm Fink Verlag.

Ellenbogen, Josh. 2008. "Camera and Mind." *Representations* 101 (1): 86–115.

Friedberg, Anne. 2006. *The Virtual Window: From Alberti to Microsoft.* Cambridge, MA: MIT Press.

Galison, Peter. 1997. *Image & Logic: A Material Culture of Microphysics.* Chicago: University of Chicago Press.

Galison, Peter, and Rob Moss. 2008. *Secrecy.* Redacted Pictures. DVD.

Gell, Alfred. 1998. *Art and Agency: An Anthropological Theory.* Oxford: Clarendon.

Hacking, Ian. 1983. *Representing and Intervening: Introductory Topics in the Philosophy of Natural Science.* Cambridge: Cambridge University Press.

Harman, Graham. 2009. *The Prince of Networks: Bruno Latour and Metaphysics.* Melbourne: Re.Press.

Harman, Graham. 2011. "On the Undermining of Objects: Grant, Bruno and Radical Philosophy." In *The Speculative Turn: Continental Materialism and Realism*, edited by Levi R. Bryant, Nick Srnicek and Graham Harman, 21–40. Melbourne: Re.Press.

Henare, Amiria, Martin Holbraad and Sari Wastell. 2007. "Introduction." In *Thinking through Things: Theorising Artefacts in Ethnographic Perspective*, edited by Amiria Henare, Martin Holbraad and Sari Wastell, 1–31. London: Routledge.

Hoel, Aud Sissel. 2011. "Thinking "Difference" Differently: Cassirer versus Derrida on Symbolic Mediation." *Synthese* 179 (1): 75–91.

Hoel, Aud Sissel. 2012. "Technics of Thinking." In *Ernst Cassirer on Form and Technology: Contemporary Readings*, edited by Aud Sissel Hoel and Ingvild Folkvord, 65–91. Basingstoke: Palgrave Macmillan.

Hoel, Aud Sissel, and Iris van der Tuin. 2013. "The Ontological Force of Technicity: Reading Cassirer and Simondon Diffractively." *Philosophy and Technology* 26 (2): 187–202.

Hogan, Pamela. 2000. *Ultimate Weapon: The H-Bomb Dilemma*. Superbomb Documentary Production Company. DVD.

Horn, Robert. 1998. *Visual Language: Global Communication for the 21st Century*. San Francisco: MacroVU.

Jones, Caroline A., ed. 2006. *Sensorium: Embodied Experience, Technology and Contemporary Art*. Cambridge, MA: MIT Press.

Jones, Caroline A., and Peter Galison, eds. 1998. *Picturing Science Producing Art*. New York: Routledge.

Kemp, Martin. 2006. *Seen | Unseen: Art, Science, and Intuition from Leonardo to the Hubble Telescope*. Oxford: Oxford University Press.

Knuuttila, Tarja, Martina Merz and Erika Mattila. 2006. "Editorial." In "Computer Models and Simulations in Scientific Practice," edited by Tarja Knuuttila, Martina Merz and Erika Mattila. Special issue, *Science Studies* 19 (1): 3–11.

Latour, Bruno, and Steve Woolgar. 1979. *Laboratory Life: The Social Construction of Scientific Facts*. Beverly Hills: SAGE.

Latour, Bruno. 2005. *Reassembling the Social: An Introduction to Actor-Network-Theory*. Oxford: Oxford University Press.

Lynch, Michael, and Steve Woolgar, eds. 1990. *Representation in Scientific Practice*. Cambridge, MA: MIT Press.

Manovich, Lev. 2001. *The Language of New Media*. Cambridge, MA: MIT Press.

McCormick, Bruce, Thomas DeFanti and Maxine Brown, eds. 1987. "Visualization in Scientific Computing." Special issue, *Computer Graphics* 21 (6).

Merz, Martina. 2006. "Locating the Dry Lab on the Lab Map." In *Simulation: Pragmatic Construction of Reality*, edited by Johannes Lenhard, Günter. Küppers and Terry Shinn, 155–172. Dordrecht: Springer.

Mitchell, W. J. T. 1994. "The Pictorial Turn." In *Picture Theory: Essays on Verbal and Visual Representation*, 11–34. Chicago: University of Chicago Press.

Neuhaus, Fabian, and Timothy Webmoor. 2012. "Agile Ethics for Massified Research and Visualisation." *Information, Communication and Society* 15 (1): 43–65.

Olsen, Bjørnar, Michael Shanks, Timothy Webmoor and Christopher Witmore. 2012. *Archaeology: The Discipline of Things*. Berkeley: University of California Press.

Pinney, Christopher. 1990. "The Quick and the Dead: Images, Time, and Truth." *Visual Anthropology Review* 6 (2): 42–54.

Pinney, Christopher. 2004. *"Photos of the Gods": The Printed Image and Political Struggle in India*. London: Reaktion Books.

Rheingold, Howard. 1992. *Virtual Reality*. New York: Simon and Schuster.

Rogers, Richard. 2009. *The End of the Virtual: Digital Methods*. Amsterdam: Vossiuspers.

Rudwick, Martin J. S. 1976. "The Emergence of a Visual Language for Geological Science 1760–1840." *History of Science* 14 (3): 149–195.

Stafford, Barbara Maria. 1991. *Body Criticism: Imaging the Unseen in Enlightenment Art and Medicine*. Cambridge, MA: MIT Press.

Stafford, Barbara Maria. 1996. *Good Looking: Essays on the Virtue of Images*. Cambridge, MA: MIT Press.

Thrift, Nigel. 2004. "Remembering the Technological Unconscious by Foregrounding Knowledges of Position." *Environment and Planning D: Society and Space* 22 (1): 175–190.

Trentmann, Frank. 2009. "Materiality in the Future of History: Things, Practices, and Politics." *Journal of British Studies* 48: 283–307.

Ware, Colin. 2000. *Information Visualization: Perception for Design*. San Francisco: Morgan Kaufman.

Webmoor, Timothy. 2013. "STS, Symmetry, Archaeology." In *The Oxford Handbook of the Archaeology of the Contemporary World*, edited by Paul Graves-Brown, Rodney Harrison and Angela Piccini, 105–120. Oxford: Oxford University Press.

Webmoor, Timothy. 2014. "Object-Oriented Metrologies of Care and the Proximate Ruin of Building 500." In *Ruin Memories: Materialities, Aesthetics and the Archaeology of the Recent Past*, edited by Bjørnar Olsen and Þóra Pétursdóttir, 462–485. London: Routledge.

West, Thomas G. 2004. *Thinking Like Einstein: Returning to Our Roots with the Emerging Revolution in Computer Information Visualization*. New York: Prometheus Books.

Woolgar, Steve, and Javier Lezaun, eds. 2013. *A Turn to Ontology?* Special issue, *Social Studies of Science* 43 (3).

Wouters, Paul, Anne Beaulieu, Andrea Scharnhorst and Sally Wyatt, eds. 2013. *Virtual Knowledge: Experimenting in the Humanities and the Social Sciences*. Cambridge, MA: MIT Press.

Yaneva, Albena. 2012. *Mapping Controversies in Architecture*. Farnham: Ashgate.

Part I

Visualization in the
Age of Computerization

1 Algorithmic Alchemy, or the Work of Code in the Age of Computerized Visualization

Timothy Webmoor

INTRODUCTION

"I'm doing something very dangerous right now" (Informant 1, July 8, 2010).

"Yah, now is not a good time for me!" (Informant 2, July 8, 2010).

Ethnographic silence can speak volumes. Despite prompts from the anthropologist, the dialogue dried up. Falling back on observation, the two informants were rapidly, if calmly, moving between their multiple program windows on their multiple computer displays. I had been observing this customary activity of coding visualizations for nearly a month now—a visual multitasking that is so characteristic of the post–Microsoft Windows age (Friedberg 2006). Looking around the open plan office setting, everyone was huddled in front of a workstation. Unlike ethnographic work in "wet labs," where the setting and activities at hand differ from the anthropologist's own cubicled site of production, this dry lab (Merz 2006) seemed so mundane and familiar, as a fellow office worker and computer user, that admittedly I had little idea of how to gain any analytic purchase as their resident anthropologist. What was interesting about what these researchers and programmers were doing?

It wasn't until I had moved on to questioning some of the PhD researchers in the cramped backroom that the importance of what the two informants were doing was overheard: "Well done! So it's live?" (Director, July 8, 2010). The two programmers had launched a web-based survey program, where visitors to the site could create structured questionnaires for other visitors to answer. The responses would then be compiled for each visitor and integrated with geo-locational information obtained from browser-based statistics to display results spatially on a map. It was part of what this laboratory was well known for: mashing up and visualizing crowd-sourced and other "open" data. While focused upon the UK, within a few months the platform had received over 25,000 visitors from around the world. The interface looked deceptively simple, even comical given a cartoon giraffe graced the splash page as a mascot of sorts. However, the reticence the day of its launch was due to the sheer labor of coding to dissimulate the complicated operations allowing such an "open" visualization. "If you're going to allow the world to ask the rest of the world anything, it is actually quite complicated" (Director, June 11, 2010). As one of the programmers later explained his shifting of attention between paper notes, notes left in the code, and the code itself, "I have to keep abreast of what I've done because an error . . . (pause) . . . altering the back channel infrastructure goes live on a website" (Informant 1, July 21, 2010).

Being familiar with basic HTML, a type of code or, more precisely, a markup text often used to render websites, I knew the two programmers were writing and debugging code that day. Alphanumeric lines, full of symbols and incongruous capitalizations and spacing, were recognizable enough; at the lab this *lingua franca* was everywhere apparent and their multiple screens were full of lines of code. Indeed, there were "1,000 lines just for the web page view itself [of the web-based survey platform]" (Informant 2, June 6, 2011)—that is, for the window on the screen to correctly size, place and frame the visualized data. Yet when looking at what they were doing there was little to see, per se. There was no large image on their screens to anchor my visual attention—a link between the programming they were engrossed in and what it was displaying, or, more accurately, the dispersed data the code was locating, compiling and rendering in the visual register.

The web-based survey and visualization platform was just one of several visualizing platforms that the laboratory was working on. For other types of completed visualizations rendering large amounts of data—for example, a London transport model based upon publicly available data from the Transport for London (TfL) authority—you might have much more coding: "This is about 6,000 lines of code, for this visualization [of London traffic]" (Informant 3, June 10, 2011) (Figure 1.1).

Over the course of roughly a year during which I regularly visited the visualization laboratory in London, I witnessed the process of launching many such web-based visualizations, some from the initial designing sessions around the whiteboards to the critical launches. A core orientation of the research lab was a desire to make the increasingly large amounts of data in digital form accessible to scholars and the interested public. Much information has become widely available from government authorities (such as the TfL) through mandates to make collected digital data publicly available, for instance, through the 2000 Freedom of Information Act in the UK, or the 1996 Electronic Freedom of Information Act Amendments in the US. Massive quantities are also being generated through our everyday engagements with the Internet. Of course, the tracking of our clicks, "likes," visited pages, search keywords, browsing patterns, and even email content has been exploited by Internet marketing and service companies since the commercialization of the Internet in the late 1990s (see Neuhaus and Webmoor 2012 on research with social media). Yet embedded within an

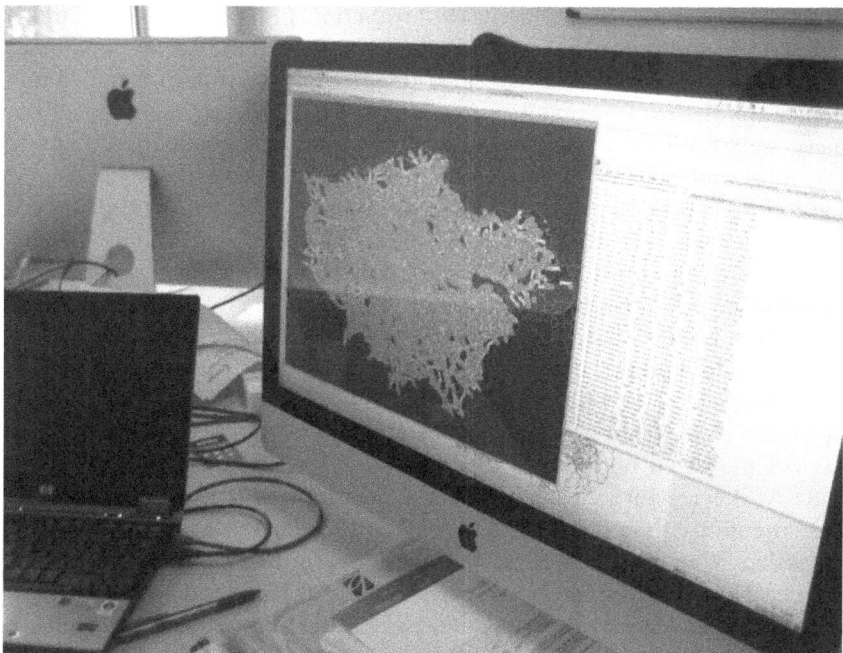

Figure 1.1 Lines of code in the programming language C++ (on right) rendering the visualization (on left) of a London transport model (Informant 3, June 10, 2011).

academic institution, this laboratory was at the "bleeding edge" of harvesting and visualizing such open databases and other traces left on the Internet for scholarly purposes. Operating in a radically new arena for potential research, there is a growing discussion in the social sciences and digital humanities over how to adequately and ethically data mine our "digital heritage" (Webmoor 2008; Bredl, Hünniger and Jensen 2012; Giglietto, Rossi and Bennato 2012). Irrespective of what sources of data were being rendered into the visual register by programmers and researchers at this lab, I constantly found myself asking for a visual counterpart to their incessant coding: "Can you show me what the code is doing?" I needed *to see* what they were up to.

CODEWORK

Code, or specifically working with code as a software programmer, has often been portrayed as a complicated and arcane activity. Broadly defined, code is "[a]ny system of symbols and rules for expressing information or instructions in a form usable by a computer or other machine for processing or transmitting information" (OED 2013). Of course, like language, there are many forms of code: C++, JavaScript, PHP, Python—to name a few more common ones discussed later. The ability to "speak computer" confers on programmers a perceived image of possessing inscrutable and potent abilities to get computers to comply.[1] Part nerd, part hero, programmers and specifically hacker culture have been celebrated in cyberpunk literature and cinema for being mysterious and libertarian.[2] The gothic sensibility informing such portrayals reinforces a darkened and distanced view of working with code.

The academic study of code, particularly from the science and technology studies (STS) perspective of exhibiting what innervates the quintessential "black boxes" that are our computing devices, has only recently been pursued ethnographically (e.g., Coleman and Golub 2008; Demazière, Horn and Zune 2007; Kelty 2008). Oftentimes, however, such studies scale out from code, from a consideration of its performative role in generating computerized outputs, to discuss the identity and social practices *of* code workers. More closely examining working with code has received less attention (though see Brooker, Greiffenhagen and Sharrock 2011; Rooksby, Martin and Rouncefield's 2006 ethnomethodological study). Sterne (2003) addresses the absent presence of code and software more generally in academic work and suggests it is due to the analytic challenge that code presents. The reasons for this relate to my own ethnographic encounter. It is boring. It is also nonindexical of visual outputs (as least to the untrained eye unfamiliar with "reading" code; see Rooksby, Martin and Rouncefield 2006). In other words, code, like Thrift's (2004) "technological unconscious," tends to recede from immediate attention into infrastructural systems sustaining and enabling topics and practices of concern. Adrian Mackenzie, in his excellent study *Cutting Code* (2006, 2), describes how software is felt to be intangible and immaterial, and for this reason it is often on the fringe of academic and commercial analyses of digital media. He is surely right to bemoan not taking code seriously, downplayed as it is in favor of supposed higher-order gestalt shifts in culture ("convergence"), political economy ("digital democracy" and "radical sharing") and globalization ("network society"). No doubt the "technical practices of programming interlace with cultural practices" (ibid., 4), with the shaping and reshaping of sociality, forms of collectivity and ideas of selfhood; what Manovich (2001, 45) termed "trans-coding" (e.g., Ghosh 2005; Himanen, Torvalds and Castells 2002; Lessig 2004; Weber 2004; for academic impacts see Bartscherer and Coover 2011). However, these larger order processes have dominated analyses of the significance involving the ubiquity of computer code.

Boring and analytically slippery, code is also highly ambiguous. The *Oxford Dictionary of Computing* (1996) offers no less than 113 technical terms that use the word "code" in the domain of computer science and information technology. So despite acknowledging the question "Why is it hard to pin down what software is?" (2006, 19), Mackenzie, a sociologist of science, admirably takes up the summons in his work. For Mackenzie, code confounds normative concepts in the humanities and social sciences. It simply does not sit still long enough to be easily assigned to conventional explanatory categories, to be labeled as object or practice, representation or signified, agent or effect, process or event. He calls this "the shifting status of code" (ibid., 18). Mackenzie's useful approach is to stitch together code with agency. Based upon Alfred Gell's (1998) innovative anthropological analysis of art and agency, Mackenzie (2006, 2005) pursues an understanding of software and code in terms of its performative capacity. "Code itself is structured as a distribution of agency" (2006, 19). To string together what he sees as distributed events involving code's agency, he takes up another anthropologist's methodological injunction to pursue "multisited ethnography" (Marcus 1995). In terms of how code is made to travel, distributed globally across information and communication technologies (ICTs) and networked servers as a mutable mobile (cf. Latour 1986), this approach permits Mackenzie to follow (the action of) code and offer one of the first non-technical considerations of its importance in the blood flow of contemporary science, commerce and society.

Given its mobility, mutability, its slippery states, code can usefully be studied through such network approaches. Yet I am sympathetic with recent moves within ethnography to reassess the importance of locality and resist the tendency (post-globalization) to scale out (see Candea 2009; Falzon 2009; in STS see Lynch 1993). While there is of course a vast infrastructural network that supports the work code performs in terms of the "final" visualizations, which will be discussed with respect to "middle-ware", most of the work involving code happens in definite local settings—in this case, in a mid-sized visualization and research laboratory in central London.

Describing how code works and what it does for the "hackers" of computerized visualizations will help ground the larger order studies of cultural impacts of computerization, as well as complement the more detailed research into the effects of computerization on scientific practices. I am, therefore, going to pass over the much studied effects of software in the workplace (e.g., Flowers 1996; Hughes and Cotterell 2002) and focus upon when technology is the work (Grint and Woolgar 1997; Hine 2006). Staying close to code entails unpacking what occurs at the multiple screens on programmers' computers. Like a summer holiday spent at home, it is mundane and a little boring to "stay local," but like the launch of the new web-based open survey visualizer that tense day, there are all the same quite complex operations taking place with code.

With the computerization of data and visualizations, the work with code weaves together many formerly distinct roles. This workflow wraps together the practices of: sourcing data to be visualized; programming to transform and render data visually; visualizing as a supposed final stage. I term these activities "codework." Merging often sequential stages involved with the generation of visual outputs, I highlight how proliferating web-based visualizations challenge analytic models oriented by paper-based media. Computerized visualizations, such as those in this case study, are open-ended. They require constant care in the form of coding in order to be sustained on the Internet. Moreover, they are open in terms of their continuing influence in a feedback cycle that plays into both the sourcing of data and the programming involved to render the data visually.

Codework, as an emergent and distinct form of practice in scientific research involving visualization, also blends several sets of binary categories often deployed in visual studies: private/public, visible/invisible, material/immaterial. While these are of interest, I focus upon the manner in which code confounds the binary of creativity/containment and discuss the implications for the political economy of similar visualization labs and the accountability of codework. For clarity, I partially parse these categories and activities in what follows.

SOURCING/PROGRAMMING/VISUALIZING

The July 6, 2010 edition of *The Guardian*, a very popular UK newspaper, featured a map of London on the second page compiled from "tweets," or posts to the microblogging service Twitter. It resembled a topographic map in that it visually depicted the density of tweeting activity around the city by using classic hypsometric shading from landscape representations. Given that there was a mountain over Soho, it was apparent to anyone familiar with the city that the data being visualized were not topographical. The caption read, "London's Twitterscape: Mapping the City Tweet by Tweet."[3] It was good publicity for the lab. The visualizations were a spin-off or side project of one of the PhD researchers. As he stated, it was an experiment "to get at the social physics of large cities through Twitter activity" (Informant 4, July 7, 2010). It was one of my earliest visits to the lab at a time when research deploying emergent social media such as Facebook, Flickr, Foursquare and Twitter, or other online sources such as Wikipedia editing activity, was in its infancy (see Viégas and Wattenberg 2004 as an early example). Indeed, many of these now popular online services did not exist before 2006.

Sourcing the data to visualize from these online platforms is not particularly difficult. It does, however, take an understanding of how the data are encoded, how they might be "mined" and made portable with appropriate programming, and whether the information will be amenable to visualizing. These programming skills are driven by a creative acumen; knowing where to look online for information relevant to research and/or commercial interests, and whether it might provide interesting and useful, or at least aesthetic, visualizations. Creative sourcing and programming are necessary crafts of codework.

Many online services, such as Twitter, provide data through an application programming interface (API). Doing so allows third-party developers to provide "bolt-on" applications, and this extensibility benefits the service provider through increased usage. "There's an app for that!": It is much like developing "apps" for Apple iTunes or Google Android. Importantly, though, sourcing these APIs or "scraping" web pages for data to visualize does require programming, and both the format of the data and the programming language(s) involved heavily determine the "final" visualizations themselves.

In the case of Twitter, the company streams "live" a whole set of information bundled with every tweet. Most of this information, such as Internet protocol (IP) location, browser or service used to tweet, link to user profile and sometimes latitude and longitude coordinates (with a 5–15 m accuracy), is not apparent to users of the service.[4] The researchers at the London lab applied to Twitter to download this open data from their development site.[5] The format is key for what types of visualization will be possible, or at least how much translation of the data by code will be required. For instance, as discussed later, many open data sources are already encoded in a spatial format like Keyhole Markup Language (KML) for display in industry standard analytic software and mapping platforms (for Google Earth or in ESRI's geographic information system [GIS] programs such as ArcGIS). Twitter, like most social media companies, government institutions and scientific organizations, formats its data as comma-separated values (CSV). For simplicity and portability across programming languages, this format for organizing data has become the primary *de facto* standard for open datasets. Information is arranged in tabular format much like an Excel or Numbers spreadsheet, and values are separated by either a comma or a tab-spacing (tabular-space values [TSV] format is a variety of CSV). The London lab logged a week's worth of tweets for various metropolises. This amounted to raw data tables containing about 150,000 tweets and over 1.5 million discrete data points for each city. Such massive datasets could be visualized based upon various criteria—for instance, semantically for word frequency or patterning. Importantly, being in CSV format means that the Twitter data are highly mutable by a wide range of programming languages, and therefore there are a number of possible paths to visualization.

The researcher was interested in tying such Twitter "landscapes" into his larger PhD research involving travel patterns in the city of London. Spatial and temporal aspects were therefore most important, and a program was written to mine the data for spatial coordinates within a certain radius of the urban centers, as well as for time stamps. When and where a Twitter user sent a tweet could then be plotted. Once aggregated, the visualizations indicated patterns of use in terms of diurnal and weekly activity. They also suggested varying usage around the cities of concern.

Sourcing data packaged as CSV files was common for the laboratory. Indeed, in step with the growing online, open data repositories, where much data are user-generated and least user-contributed, the lab was developing a visualizing middle-ware program to display spatial data based upon Open-StreetMap and OpenLayers.[6] "A place to share maps and compare data visually," as their website states, is the goal. Unlike either large commercial companies that typically fund and manage server farms where these large datasets are uploaded, such as Google's Spreadsheets or Yahoo!'s Pipes, or commercially funded research groups like IBM's ManyEyes,[7] this lab was providing a visualizing tool, or "visualizer," that would fetch data which was located remotely. Otherwise, despite the recent purchase of three new Dell stacked servers located in a closet in the hallway, there simply was not enough storage space for the modest lab to host the huge datasets. Described as "a service for researchers," the web-based platform would, for example, "serve up on an ad hoc basis a visualization of the dataset whenever a query was made by a user" (Informant 1, November 10, 2010). As he went on to unpack the operation, when a user selected a set of statistics (e.g., crime statistics) to display over a selected region (e.g., the UK), there were a host of operations that had to transpire rapidly, and all hinged upon the web-based platform and the coding the researcher had written and "debugged" (and rewritten). These operations might be thought of as putting together layers consisting of different information and different formats in order to build a complexly laminar visualization. As the informant described, the programming allows the "fusing of CSV data [like census data] with geo-spatial data [coordinates] on a tile by tile basis [portions of the map viewed on the screen] . . . this involves three servers and six websites" (Informant 1, November 10, 2010). Data, plug-ins and the program code itself were variously dispersed and are pulled together to form the visualization on a user's screen—hence the "middle" or go-between function of the software package developed by the lab and later released as GMapCreator. For programming web-based visualizations, code is coordination work.

In addition to the growing online datasets stored primarily as CSV files, many government agencies now provide their stored data in accessible digital repositories. The UK's data.london.gov.uk was another popular archive where the visualization lab and other researchers were obtaining UK census data. This was variously fetched and displayed as a dynamic, interactive map through the lab's web-based platform. Twenty-seven months after the initial launch of the visualizing platform, 885 datasets had been "uploaded" to the map visualizer (or more precisely shared through linking to a remote server where the datasets were physically stored) and there were "about 10,000 users for the program" (Informant 1, November 10, 2010).

Other data was being sourced by the lab through scraping. While the "ready-made" data stream from Twitter's API or other online data repository requires some initial coding to obtain the data—for instance, writing a program to log data for a specified period of time as in the case of the Twitterscapes—web-scraping typically requires more involved code-work. Frequently, the data scraped must go through a series of steps all defined by code in order to be useful; and in many cases converted into a standard data format such as CSV.

Let's consider another example from the visualization lab. One of the programmers was interested in getting at "the social demography of, and city dynamics relating to, bike share usage" (Informant 5, October 19, 2010). Bike share or bike rentals programs had recently become quite popular in international cities. London's own bike sharing scheme, known as the Barclay's Cycle Hire after its principal commercial contributor, was launched on June 30, 2010, and represents a quite large and well-utilized example. Visualizing and making publicly accessible the status of the bike share schemes piggybacks off of the data that the corporate managers of the various schemes collect for operational purposes. The number of bikes at each docking station (London has over 560 stations) is updated electronically every three minutes. The stages involved begin with the programmer using the "view source" feature in the web browser Firefox to de-visualize the "front end" of the commercial websites in order to assess what types of information are encoded. The "back end" or source code of the scheme's website showed several types of information that could be rendered spatially. Specifically, time stamps, dock location and number of bicycles were data he thought could be harvested from these websites. With visualizations in mind, he was confident "that they would be something cool to put out there" (Informant 5, September 23, 2010). He wrote a code in Python to parse the information scraped to narrow in on geographic coordinates and number of bicycles. Using a MySQL database to store the information, the Python program pulled the selected data into CSV format by removing extraneous information (such as HTML markup). A cron program was written to schedule how frequently the web scraping takes place. Finally, the programmer aggregated the information scraped for each individual bike station to scale up to the entire system or city. To visualize the data, he used JavaScript to make it compatible with many web-based map displays, such as Google Maps. In this case, he used OpenStreetMap with (near) real time information displayed for ninety-nine cities worldwide (Figure 1.2). Finally, he used Google's Visualization API[8] to generate and embed the charts and graphs. Several months after the world bike sharing schemes visualizer went live, the programmer related how he was contacted by the corporate managers of four different international cities asking him "to take down the visualizations . . . likely because it made apparent which schemes were being underutilized and poorly managed" (Informant 7, October 27, 2010).

Figure 1.2 Cached MySQL database consisting of number of bicycles and time stamps "scraped" from websites (on left) with the near real time visualization of the data in JavaScript (on right; in this case London's bicycle share scheme). Bubbles indicate location of bicycle docks, with size proportional to the number of bicycles currently available. Colored lines show movement of bicycles between dock stations (Informant 5, October 19, 2010).

SOURCING/*PROGRAMMING*/VISUALIZING

Visualizations crafted by the lab were clearly having a form of public "impact" and response. While not necessarily the type of impact accredited within the political economy of academia, the lab was all the same successful in attracting more funding based upon a reputation of expanding public engagement with its web presence. As the director remarked, "We are flush with grant money right now . . . we are chasing the impact" (October 19, 2010).

Not dissimilarly to other studies of scientific visualizations (e.g., Beaulieu 2001; Lynch and Edgerton 1988; Lynch and Ridder-Vignone this volume), there was a need to pursue public outreach achieved through accessible visualizations. At the same time, the researchers and programmers themselves de-emphasized such "final" visualizations. Of course, I overheard several informants speaking approvingly of the visual styles used by colleagues in the lab. "He has good visualizations" (Informant 7, October 27, 2010) was a typical, if infrequent, response to my queries about their fellow researchers. More often, though, were the comments:

"He has clean code" (Informant 5, October 27, 2010); or "the style of programming is very important . . . everything around the actual algorithm, even the commenting [leaving lines within the code prefaced by '//' so as not to be readable/executable by the computer] is very important" (Informant 2, June 10, 2011). More than visualizations or data, the programming skill that made both possible was identified as the guarantor of reliability in web-based visualizations.

"Code is king" assertions contrast with previous studies where researchers engaging visualizations stated greater confidence in, or preference for, the supposed raw data visualized. In this laboratory, the programmers and researchers tended to hold a conflicted opinion of data. Ingenuity in finding sources of data to visualize was admired, such as with the bike share scheme or the Twitter maps. Equally, the acts of researchers who leveraged open data initiatives to amass new repositories, such as making requests through the Freedom of Information Act to the Transport for London (TfL) to release transportation statistics, were approvingly mentioned (e.g., Informant 5, November 10, 2010). Yet once sourced, the data tended to recede into the background or bedrock as a substrate to be worked upon through programming skill.

> Everyone has their preferred way of programming, and preferred style of programming . . . The data is much more clearly defined. The structure of the data and how you interface, or interact, with it. Whereas programming is so much more complex. It's not so easy . . . maybe sounds like it's isolated. But it's really hundreds of lines of code. (Informant 3, June 10, 2011)

Like this informant, many felt that the data were fairly "static and closed," and for this reason were less problematic and, consequently, less interesting to work with. One programmer explained, with reference to the prevalence of digital data in CSV format, that you need to "use a base-set that was ubiquitous . . . because web-based visualizations were changing constantly . . . with new versions being released every week or so" (Informant 1, June 13, 2011). Such a perception of data as a relatively stable "base" was often declared in contrast to the code to "mash up" the data and render it readable by the many changing platforms and plugins. As a go-between, code has to be dynamic to ensure the data remains compatible with the perpetually changing platforms for visualizing. Programmers often bemoaned how much time they spent updating their code to keep their visualizations and platforms live on the Internet. Informant 1 (November 10, 2010) explained how he was "developing a point-click interface to make it easier to use the site [the mapping visualizer that the lab hosted], but it requires much more KML [a standard geospatial markup language] to make it compatible with Google Maps and OpenStreetMap [which the site used to display the maps]." When Google or another API provider updated their software, the programmers often had to update the code for their visualizations accordingly.

Perhaps unsurprisingly, activities that required skill were more highly regarded. These preferences fed into the treatment of data vis-à-vis code in terms of lab management. There was much informal sharing of datasets that had already been sourced, sharing links to where relevant data might be found, or otherwise discussing interesting and pertinent information for one another's research projects. To make sharing and collaborating with data easier and more accountable, the lab was beginning to set up a data repository using Trac, an open source project management program, on a central server. When asked about a code repository, the informant identified the need, but admitted that a "data repository was much less of a problem" (Informant 5, October 13, 2010). Instead, despite "revision control being essential to programming . . . and taught in the 2nd year of software engineering," it was largely left to the individual programmer to create a "code repository so that [you] can back up to where [the software] was working . . . like a safety net" (Informant 2, March 16, 2011).

Updating the felicitous phrase describing Victorian sex, code was everywhere admired, spoken about and inherent to the survival of the lab, but never shared. This incongruous observation prompted a conversation later in the fieldwork:

TW: You don't share code, you share data?
Informant: Mostly data, not so much code. The thing with data, once you
          load it and you have your algorithms, you can start to transform
          the data in different ways. You can also (pause) . . . quite often
          you download data in a certain format, and you load it, and your
          programming transforms it in different ways. Then it becomes
          useful . . . With C++ there is no boundary to what you can do.
TW: So you can do whatever with the data. So you do that by writing
          new code? Do you write a bit of code in C++?
Informant: I do it all the time. It's how I spend my days. (Informant 3, June
          10, 2011)

Programming was held in deference for two principal reasons. First, as a creative skill it was seen as proprietary (see Graham 2004 on the parallels between artists and programmers). It was not something that could be straightforwardly taught, but as a craft it was learned on the job; and programmers strove to improve this skill in order to write the cleanest, most minimalist code. This fostered a definite peer awareness and review of code. These interpersonal dynamics fed into the larger management and economy of the lab already mentioned. This lab was chasing the impact. To do so they were creating "fast visualizations." In addition, as discussed with respect to the mapping visualizer, web-based visualizations rely upon middle-ware or go-between programs gathering distributed data and rendering it on an ad hoc basis on the computer screen. Given the larger infrastructural medium, the lab's visualizations needed constant coding and recoding in order to maintain operability with the rapidly changing software platforms and APIs that they were tied to. Each researcher who began a project had to quickly and on an ad hoc basis develop code that could render source data visually. Intended for rapid results but not necessarily long-term sustainability, the only reasonable way to manage such projects was to minimize confusion and the potential for working at cross-purposes by leaving the coding to individuals. At a smaller research lab where programming happens, this was feasible. At larger, corporate laboratories, the many tasks of codework are often broken up in a Fordist manner among several specialists: software engineers, network engineers, graphic designers. All of whom may be involved in creating web-based visualizing platforms. Bolted together by teams, such coding is more standardized, minimalist and therefore more compatible and fungible.

In contrast, the London lab found itself in a double-bind of sorts. They clearly knew they needed some form of accountability of the code produced by the lab: "Revision control is essential for the lab" (Informant 2, March 2, 2011). At the same time, the success of the lab in garnering financial support, the "soft funding" it was entirely dependent upon, was largely due "to the creative mix of code and tools" and the "hands-off management style" (Director, March 2, 2011). Individual researchers were therefore largely left to practice their own codework.

Secondly, in addition to being a highly creative and skilled craft, writing code, and specifically the code itself, was seen as dynamic and potent in relation to data. Mackenzie discusses the agency of code in general terms with respect to its ability to have effects at a distance (2005, 2006)—for instance, forging a shared sense of identity among the internationally based software engineers who develop the open source software Unix. For this lab, the software they were writing had agency in the very definite sense of transforming data. You want to "build an algorithm so you have good performance" (Informant 6, October 13, 2010). Unlike code more generally, which includes human-readable comments and other instructions, an algorithm is a subset or specific type of code that is expressly written for a computer to perform a specified function in a defined manner. The defining function of many algorithms in computer programming is the ability to manipulate data. For instance, as the informant describes ahead, a basic operation of algorithms is precisely the transformation of data from one format (e.g., the markup language KML) to a different format (e.g., C++).

An algorithm performs a well-defined task. Like sorting a series of numbers, for example. Like a starting point [for the] input of data and output of data in different formats. You can do everything with source code. You can write algorithms, but you can do all other kinds of stuff as well. (Informant 3, June 10, 2011)

Code as transitive revealed the conflicting view of data as being less than stable. In fact, during many conversations, code's potency was estimated in fairly direct relation to how malleable it rendered data.

TW: You transform these two sources of data that you have into XML files, so that you can look at them in ArcGIS?

Informant: From ArcGIS you understand how the data, how these things are supposed to be connected. So then you can connect the network in the way it's supposed to be connected, in your connection structure that you have defined in your code, in C++.

TW: Does it have to be in C++, or is that a standard for running simulations? . . . So you have to transform this data in C++, let me get at this, for a couple of reasons: One, you are familiar with it; two, it's going to be much faster when you make queries [as opposed to ArcGIS]. Anything else?

Informant: Well, I'm familiar with the code. (Informant 6, October 13, 2010)

Code's purpose was to translate and to work with data. Yet different types of code worked on data differently. Part of this was due to personal preference and background. What certain programmers could do with code depended upon their familiarity with it. For this reason, depending upon the programmer, certain code was asserted to be "top end" or "higher order." This implies that the coding language is more generalized and so could be written to perform many types of tasks. It also means, however, that the code must be programmed much more extensively. Whatever code was preferred, the acknowledgment that it transformed and manipulated data rarely led to discussion of a potential corollary: that data were somehow "constructed" or may become flawed. Part of this has to do with the data asserting a measure of independence from code. More specifially, the format of the data partially determines the type of visualization pursued and so constrains to a certain degree the coding deployed. More emphasis was, however, given to code's neutral operation upon data. It was held to merely translate the data's format for "readability" across the networked computers and programs in order to render the visualization. Put another way, code transforms metadata not data. The view that code transformed without corruption became most apparent when researchers discussed the feedback role of visualization in examining and correcting the original datasets.

SOURCING/PROGRAMMING/*VISUALIZING*

Most visualizations mashed up by the lab were not finalized outputs. Just as the format of the sourced data influenced the programming required and the choices for visualization, the visualizations themselves recursively looped back into this process of codework. Several informants flipped the usual expectation that visualizations were end products in the information chain by discussing their integral role at the beginning of the process.

> It's a necessary first step to try and visualize certain facets of the information. Because it does give you a really quick way of orienting your research. You can see certain patterns straightaway . . . you literally do need to see the big picture sometimes. It informed my research completely . . . There are some themes in your research. But you still, (pause) the visualization informs your trajectory . . . Is part of the trajectory of your research . . . You draw on that . . . because you see something. (Informant 8, October 27, 2010)

For this informant, deploying "sample" visualizations allowed him to identify patterns or other salient details in an otherwise enormous corpus of data. He was researching with Foursquare, a social media service that personalizes information based upon location. As he stated, he was "mining their API to harvest 300,000,000 records" (Informant 8, October 27, 2010). Awash in data, he needed to reduce the complexity in order to identify and anchor research problems. Much like the lab's resident anthropologist, he needed a visual counterpart to what was otherwise undifferentiated and unfamiliar clutter on the computer screen.

More than suggesting what to do with mined data, another researcher noted with pride that the working visualizations actually identified flaws in the data. He was coding with open data from Transport for London and the UK Ordnance Survey. Much of this entails merging traffic flow volume (TfL data) with geospatial coordinates (Ordnance data) to create large visuals consisting of lines (roadways, paths) and nodes (intersections). Responding to a question about accuracy in the original data and worries about creating errors through transforming the CSV files into C++, he discussed the "bridge problem." This was an instance where the visualization he began to run on the small scale actually pinpointed an error in the data. Running simulations of a small set of nodes and lines, he visually noticed traffic moving across where no roadway had been labeled. After inspecting a topographic map, he concluded that the Ordnance Survey had not plotted an overpass where it should have been.

TW: When you say you know exactly what is going on with the algorithm, does that mean you can visualize each particular node that is involved in this network?

Informant: Yes, once you program you can debug it where you can stop where the algorithm is running, you can stop it at anytime and see what kind of data it is reading and processing . . . you can monitor it, how it behaves, what it is doing. And once you check that and you are happy with that, you go on to a larger scale

> . . . once you see what it is giving you. (Informant 6, October 13, 2010)

Running visualizations allowed the researcher to feel confident in the reliability of his data as he aggregated (eventually) to the scale of the UK.

Whether visualizations were used by the lab researchers at the beginning of the process of codework to orient their investigations, or throughout the process to periodically corroborate the data, visualizations as final outputs were sometimes expected. Where this was the case, especially with respect to paper-based visualizations, many at the lab were resigned to their necessity but skeptical of the quality. Several felt it was an inappropriate medium for what were developed to be web-based and dynamic. Yet they openly acknowledged the need for such publication within the confines of academia's political economy.

> Online publishing has to be critical. For me, my research is mainly online. And that's a real problem as a PhD student. Getting published is important. And the more you publish online independently the less scope you have to publish papers . . . still my motivation is to publish online, particularly when it's dynamic . . . You might have a very, very interesting dynamic visualization that reveals a lot and its impact in the academic community might be limited . . . the main constraint is that these are printed quite small, that's why I have to kind of tweak the visuals. So they make sense when they are reproduced so small. Because I look at them at twenty inches, on a huge screen. That's the main difference, really. There's a lot of fine-grained stuff in there. (Informant 8, October 27, 2010)

Set within academic strictures of both promotion and funding, the lab's researchers found themselves needing to generate fast visualizations, while at the same time "freezing" them, or translating and reducing them, to fit traditional print.

CONCLUDING DISCUSSION: CODEWORK AND RIGHT WRITING IN SCIENTIFIC PRODUCTION

Given the way codework weaves together the many activities happening at the visualization lab, the demands of academic publication to assign definite credit became an arena for contestation. This is because programming, as a mode of writing and a skill integral to all of these activities, abrades against a tradition of hierarchical assignation of authorship going back to pre-Modern science (Shapin 1989). This tradition would restrict the role of software programming along the lines of Shapin's "invisible technicians." Writing code is not the right type of writing.

The nonproprietary attitude on the part of the programmers-researchers at the lab toward the visualized results encouraged "nonauthorship". Yet, as their resident anthropologist, I myself violated the lab's mode of production which regarded visualizations as collective achievements. I became entangled in their web of work. While, as previously mentioned, the complexities of code involved in sourcing, programing and visualizing were partially contained by leaving this entire process up to individual researchers, the lack of more defined, sequential steps involved with web-based visualizations meant that there were sometimes conflicts over authorship. Sourcing of data, as discussed, involved creativity in locating data and skill in harvesting the information through programming. These skills were bound together as part of the craft of codework. Within a collaborative environment, where researchers shared ideas for data sources, the question was raised whether a single individual ought to retain sole authorship for a visualization. This was especially so when the idea for the data mining came from another researcher. With a particular article that I co-authored at the lab, there was acrimony that the individual who identified the data source was not granted co-authorship. Additionally, another researcher helped with the programming needed to harvest the information, but was subsequently little involved in the project. Again, because of the continuity of codework, the programmer felt he ought to have been listed as a co-author.

This personal vignette occurred toward the end of my year at the lab and it suggested that I was perhaps getting too close and "going native". Not simply because I was co-authoring "outputs" of the lab as a participant, but also because of the palpable and expressed sense of altering the working dynamics of the programmers. My involvement contentiously and publicly posed the question of whether the attribution of credit at the lab, and in academia more generally, operated according to anachronous guidelines with regard to such visualizations. Might such dry labs follow the precedent of wet labs, where large teams federate their efforts together under the directorship of a senior scientist? This translates, not without contention, into a lead author with many possible secondary authors listed. However, the specific activity of codework aligned most of the lab's researchers with the so-called "hacker ethic" mode of production. That is, according to Himanen, Torvalds and Castells (2002, 63) "the open model in which the hacker gives his or her creation freely for others to use, test, and develop further." Such an ideal of radical sharing is difficult to square with the constraints of academic credit and promotion. My own awkward intervention in the lab's political economy demonstrated the tendency to relegate writing software to the invisible spectrum of scientific production.

Clearly the medium of live, web-based visualizations questions our still analog-oriented modes of credit in the academy. Ideas of web presence and public outreach through engagement with Internet-based research are not radically new. More strongly, however, as the informant working with Foursquare intimated, is that the gold standard of print publication is not an appropriate medium for this type of interactive imagery. Notions of "authorship" may not be straightforwardly applicable. Scientific production in wet-labs, such as the classic examples studied in STS (Knorr-Cetina and Mulkay 1983; Latour 1987; Latour and Woolgar 1986; Lynch 1985), disclosed the messiness of making science work. The work of code in the age of computerization further blurs the roles and stages of scientific research.

The dynamic and open visualizations depend upon middle-ware, or specifically written code, to fetch distributed sources of data and software plug-ins (with APIs). Given that the production and hosting of these datasets and plug-ins involve multiple commercial, governmental and academic sources dispersed across the Internet, the code to federate and sustain web-based visualizations must be constantly cared for (on matters of care, see Puig de la Bellacasa 2011). As a woven set of activities, I would argue that the study of visualizations, particularly the proliferation of web-based visualizations, must de facto consider the entire workflow centered upon coding. Path dependency is involved. However, it is not a linear or sequential workflow as this metaphor suggests, with each stage or activity building stepwise toward a final (or at least finalized) output. It is rather like reiterative knot making, with preferred programming languages and skills constraining both what types of data are sourced and the forms of visualization that are possible, and also the format of the data acting to constrain the possible visualizations and what programming languages are deployed. Additionally, the visualizations themselves are often used to ground-truth or check the sourced data. The output or visualization itself is held only provisionally like a knot before it is redone, adjusted, updated or taken down. As an emergent craft in the making of scientific visualizations, codework presents challenges to our analytic categories for studying science. Yet it promises to pose interesting questions for future practices in the academy.

NOTES

1. As I complete the draft of this chapter, I wait hoping that the computer science students are miraculously able to resuscitate my computer's crashed hard drive.
2. Novels by William Gibson, Neal Stephenson and Bruce Sterling set the genre of cyberpunk. Popular films such as *Blade Runner* or the *Matrix* trilogy are the genre's more well-known cinematic counterparts.
3. See the archived story at: http://www.guardian.co.uk/news/datablog/2010/jul/06/twitter-maps?INTCMP=SRCH. Last accessed 4 November, 2013.
4. These data points include: TwitterPost (actual message); TwitterID (in addition to the username every user is assigned an internal number); dateT (date and time when the message was sent); name (user or screen name); link (link to online location of this tweet); usage; Twittergeo; lang (language); profile (link to online user profile); google_location (location of user derived as a geocoded place via the user profile location); atom_content (message in atom

format); source (platform used to send the tweet); Lat (latitude) and Lon (longitude).

5. Located at: dev.twitter.com.
6. These are open source alternatives to Google Maps; see http://www.open-layers.org/ and http://www.openstreetmap.org/. Last accessed 4 November, 2013.
7. See http://drive.google.com, http://pipes.yahoo.com/pipes/, and http://www-958.ibm.com/software/analytics/manyeyes/. Last accessed 4 November, 2013.
8. A set of visualizing tools offered freely by Google: https://developers.google.com/. Last accessed 4 November, 2013.

REFERENCES

Bartscherer, Thomas, and Roderick Coover, eds. 2011. *Switching Codes: Thinking through Digital Technology in the Humanities and the Arts*. Chicago: University of Chicago Press.

Beaulieu, Anne. 2001. "Voxels in the Brain: Neuroscience, Informatics and Changing Notions of Objectivity." *Social Studies of Science* 31 (5): 635–680.

Bredl, Klaus, Julia Hünniger and Jakob Linaa Jensen. 2012. "Methods for Analyzing Social Media: Introduction to the Special Issue." *Journal of Technology in Human Services* 30: 141–144.

Brooker, Christian Greiffenhagen, and Wes Sharrock. 2011. "Scientific Findings." Paper presented at the Visualisation in the Age of Computerisation conference, March, Oxford, UK.

Candea, Matei. 2009. "Arbitrary Locations: In Defense of the Bounded Field-Site." In *Multi-Sited Ethnography: Theory, Praxis and Locality in Contemporary Research*, edited by Mark-Anthony Falzon, 25–46. Farnham: Ashgate.

Coleman, Gabriella, and Alex Golub. 2008. "Hacker Practice: Moral Genres and the Cultural Articulation of Liberalism." *Anthropological Theory* 8 (3): 255–277.

Demazière, Didier, François Horn and Marc Zune. 2007. "The Functioning of a Free Software Community: Entanglement of Three Regulation Modes—Control, Autonomous and Distributed." *Science Studies* 20 (2): 34–54.

Falzon, Mark-Anthony, ed. 2009. *Multi-sited Ethnography: Theory, Praxis and Locality in Contemporary Research*. Farnham: Ashgate.

Flowers, Stephen. 1996. *Software Failure: Management Failure—Amazing Stories and Cautionary Tales*. New York: John Wiley and Sons.

Friedberg, Anne. 2006. *The Virtual Window: From Alberti to Microsoft*. Cambridge, MA: MIT Press.

Gell, Alfred. 1998. *Art and Agency: An Anthropological Theory*. Oxford: Clarendon.

Ghosh, Rishab Aiyer, ed. 2005. *CODE: Collaborative Ownership and the Digital Economy*. Cambridge, MA: MIT Press.

Giglietto, Fabio, Luca Rossi and Davide Bennato. 2012. "The Open Laboratory: Limits and Possibilities of Using Facebook, Twitter, and YouTube as a Research Data Source." *Journal of Technology in Human Services* 30: 145–159.

Graham, P. 2004. *Hackers and Painters: Big Ideas from the Computer Age*. Sebastopol, CA: O'Reilly Press.

Grint, Keith, and Steve Woolgar. 1997. *The Machine at Work*. Cambridge: Polity Press.

Himanen, Pekka, Linus Torvalds and Manuel Castells. 2002. *The Hacker Ethic: A Radical Approach to the Philosophy of Business*. New York: Random House.

Hine, Christine. 2006. "Databases as Scientific Instruments and Their Role in the Ordering of Scientific Work." *Social Studies of Science* 36: 269–298.

Hughes, Bob, and Mike Cotterell. 2002. *Software Project Management*. 3rd ed. London: McGraw-Hill.

Kelty, Christopher. 2008. *Two Bits: The Cultural Significance of Free Software*. Durham, NC: Duke University Press.

Knorr-Cetina, Karin, and Michael Mulkay, eds. 1983. *Science Observed: Perspectives on the Social Study of Science*. London: SAGE.

Latour, Bruno. 1986. "Visualization and Cognition: Thinking with Eyes and Hands." In *Knowledge and Society: Studies in the Sociology of Culture Past and Present*, edited by Elizabeth Long and Henrika Kuklick, 1–40. London: JAI Press.

Latour, Bruno. 1987. *Science in Action: How to Follow Scientists and Engineers through Society*. Cambridge, MA: Harvard University Press.

Latour, Bruno, and Steve Woolgar. 1986. *Laboratory Life: The Construction of Scientific Facts*. 2nd ed. Princeton: Princeton University Press.

Lessig, Lawrence. 2004. *Free Culture: The Nature and Future of Creativity*. New York: Penguin Books.

Lynch, Michael. 1985. *Art and Artifact in Laboratory Science: A Study of Shop Work and Shop Talk in a Research Laboratory*. London: Routledge and Kegan Paul.

Lynch, Michael. 1993. *Scientific Practice and Ordinary Action: Ethnomethodology and Social Studies of Science*. Cambridge: Cambridge University Press.

Lynch, Michael, and Samuel Edgerton. 1988. "Aesthetics and Digital Image Processing: Representational Craft in Contemporary Astronomy." In *Picturing Power: Visual Depiction and Social Relations*, edited by Gordon Fyfe and John Law, 184–220. London: Routledge.

Mackenzie, Adrian. 2005. "The Performativity of Code: Software and Cultures of Circulation." *Theory, Culture and Society* 22 (1): 71–92.

Mackenzie, Adrian. 2006. *Cutting Code: Software and Sociality*. New York: Peter Lang.

Manovich, Lev. 2001. *The Language of New Media*. Cambridge, MA: MIT Press.

Marcus, George. 1995. "Ethnography in/of the World System: The Emergence of Multi-sited Ethnography." *Annual Review of Anthropology* 24: 95–117.

Merz, Martina. 2006. "Locating the Dry Lab on the Lab Map." In *Simulation: Pragmatic Construction of Reality*, edited by Johannes Lenhard, Günter Küppers and Terry Shinn, 155–172. Dordrecht: Springer.

Neuhaus, Fabian, and Timothy Webmoor. 2012. "Agile Ethics for Massified Research and Visualisation." *Information, Communication and Society* 15 (1): 43–65.

Oxford Dictionary of Computing. 1996. Oxford: Oxford University Press. Oxford Reference Online. Accessed May 20, 2013. http://www.oxfordreference.com/views/ENTRY.html?subview=Main&entry=t11.e4940.

Oxford English Dictionary (OED). 2013. Accessed May 24, 2013. http://oed.com.

Puig de la Bellacasa, Maria. 2011. "Matters of Care in Technoscience: Assembling Neglected Things." *Social Studies of Science* 41 (1): 85–106.

Rooksby, J., D. Martin and M. Rouncefield. 2006. "Reading as a Part of Computer Programming. An Ethnomethodological Inquiry." Unpublished ePrint. Accessed June 20, 2013. http://eprints.lancs.ac.uk/12829/.

Shapin, Steven. 1989. "The Invisible Technician." *American Scientist* 77: 554–63.

Sterne, J. 2003. "Bourdieu, Technique and Technology." *Cultural Studies* 17 (3–4): 367–389.

Thrift, Nigel. 2004. "Remembering the Technological Unconscious by Fore-grounding Knowledges of Position." *Environment & Planning D: Society & Space* 22 (1): 175–191.

Viégas, Fernanda, Martin Wattenberg and K. Dave. 2004. "Studying Cooperation and Conflict between Authors with 'History Flow' Visualizations." *Computer-Human Interaction* 6 (1): 575–582.

Webmoor, Timothy. 2008. "From Silicon Valley to the Valley of Teotihuacan: The Yahoo!©s of New Media and Digital Heritage." *Visual Anthropology Review* 24 (2): 183–200.

Weber, Steve. 2004. *The Success of Open Source.* Cambridge, MA: Harvard University Press.

2 From Spade-Work to Screen-Work
New Forms of Archaeological Discovery in Digital Space

Matt Edgeworth

INTRODUCTION

This chapter is about the influence of computers on archaeological modes of perception. It tracks changes that have taken place over the last two decades in ways that archaeologists perceive and represent material evidence. It is also about my own realization, as an archaeological practitioner as well as an ethnographer of archaeological practice, of the importance of computerization in the processes of discovery. Twenty three years ago, in an early ethnography of an excavation, I took the "act of discovery" to be located in practical encounters between archaeologists and materials out on site, mediated by craft tools such as trowel and spade. Now it is clear that the very notion of discovery needs to be rethought and perhaps even relocated to take account of the proliferation of digital devices, with screen-work as well spade-work central to the emergence and interpretation of new evidence. In speaking of the "craft" of archaeology today, the roles of computers need to be considered alongside those of more traditional tools: Indeed, this chapter explores the extent to which computers have rapidly instated themselves as embedded material and cognitive components of the craft tradition of archaeological knowledge production. Through the use of an ethnographic case study of an archaeologist searching for hidden artificial waterways on Google Earth, I compare nondigital with digital forms of discovery, and look at ways in which these are increasingly interconnected in contemporary practice.

SHIFTING ASSEMBLAGES

Although computers are the main focus of attention here, archaeology tends to be associated first and foremost with more ancient material culture—artifacts like coins, worked flints and pottery vessels, for example, or features like postholes, pits, ditches and wall-foundations uncovered in the ground. These have been created and used by human beings in other cultures, other times, to be rediscovered and reappropriated by archaeologists.

A useful archaeological term to refer to here is that of "assemblage," which can be defined as a "group of artifacts recurring together at a particular time and place" (Renfrew and Bahn 2008, 578).

As well as assemblages of items of material culture from the past, however, we also have to consider involvement of our own artifacts in the production and reproduction of archaeological knowledge. The material practice of excavation, for instance, has its own characteristic set of tools—a fairly basic technology of spades, picks, trowels, hand shovels, buckets, wheelbarrows and so on—that are used to bring archaeological evidence to light. Sometimes these are used in conjunction with heavy machines such as earthmovers and dump trucks. Equipment taken out to site also includes numerous items of recording and surveying equipment—nails, string, tapes, spirit-levels, files, pens, pencils, planning boards, planning-grids, etc.

These myriad things embody spatial ways of seeing and temporal rhythms of archaeological practice and help to mediate archaeologists' perception of unfolding material evidence, and thus form part of assemblages in the much wider sense of the term as deployed by Deleuze and Guattari (1987), Latour (1993) and Bennett (2010), denoting hybrid mixtures of human and nonhuman forces, flows and resistances, ideas and substances at work together. In the case of archaeological practice, such assemblages necessarily entail a coming together of (among other things) ancient and contemporary items of material culture.

In 1989–1990 I carried out an ethnographic study of the archaeological excavation of a Bronze Age site in Brightlingsea, UK, which examined precisely the question of how tools mediate the perceptual transactions with unfolding material evidence, at once shaping and being shaped by the encounter (Edgeworth 1991, 2003). The main focus of that work was on what I called the "act of discovery," which was defined as "the (temporal) relation between an (embodied) subject and an (emerging) object, mediated through (the use of) tools" (1991, 2). That entailed an interest both in the artifacts and features found in the ground (arrowheads, pottery shards, features, etc.) and in the tools used in the unearthing of them (trowels, spades, picks, etc.), as well as those items of kit deployed in recording and planning (grids, planning frames, tape measures, etc.). I had no qualms at that time about seeing archaeological excavation as the main site of archaeological discovery, and situating my ethnography accordingly "out in the field."

But that was before digital technology had made significant inroads into the profession, or become truly embedded as part of archaeological assemblages (in the wider Latourian sense outlined earlier, rather than the more limited archaeological meaning of the term). Computers were of course utilized by archaeologists at that time, but their use was more limited compared to today, being mainly the preserve of specialists and principally applied to the statistical analysis of archaeological data far removed from the site of discovery (see Cowgill 1967 for an early statement of the use of computers in archaeology).

Now, in the more developed digital age, the use and influence of computers are so widespread that it is not quite so clear exactly where the main site of archaeological discovery is located in space. New forms of discovery have arisen on-screen in addition to those that have always taken place out on-site. Indeed, the term "site of discovery" might usefully be taken to refer to virtual on-screen realities as well as off-screen ones. Certainly if one were to carry out an ethnographic study of archaeological discovery today, one would surely have to take a multisited approach, along the lines suggested by Wittel (2000), looking at screen-work as well as spade-work. Multisited ethnography implies the following of a research problem through different sites, looking at movements of people, ideas and material culture across space and time rather than confined to specific spatial and temporal locations (Marcus 1995; Gupta and Ferguson 1997; Falzon 2009).

In reflecting on the developments over the last two decades, it will be useful to refer back to my 1991 ethnography now and again. It gives a reasonably accurate picture of what archaeological discovery was like on a British excavation back then, and to a certain extent what it is still like today. For various reasons, excavation has remained relatively, though not completely, untouched by computerization—at least with regard to its basic perceptual encounters with materials. Before discussing the reasons for that, however, we must first take stock of recent material transformations in the working environment of archaeologists. In the following account, I will refer mainly to British commercial archaeological units where I now chiefly work, though similar transformations have been wrought in many other archaeological organizations worldwide, including academic departments and museums.

MATERIAL TRANSFORMATION OF ARCHAEOLOGICAL WORKPLACES

The material culture of archaeological offices has radically changed since I carried out my ethnography of an archaeological site. Take the technology of map-work (Webmoor 2005), for example. Gone are the inked-in plans, written matrices, paper drafts of publication drawings and so on. Gone too are the colored pencils, sets of technical pens, tracing paper pads, rulers and associated material accoutrements that once covered the desks of archaeological units. All these have now been largely replaced by computer software programs (AutoCad, Illustrator, ArcGIS, etc.). Heavy metal cabinets for hanging large maps and site plans still exist but are increasingly redundant as images are digitized and made available online.

Tasks that formerly took place at a table, writing desk or drawing board have mostly been transferred onto keyboard and monitor, with corresponding shifts taking place in the skills, postures and talk of the archaeological workplace. Networked individual workstations have largely replaced large desk-spaces for collaborative working, leading to the wholesale reorganization of working spaces. E-mail has taken over many of the functions of posted letters. Much research, marketing and publication of work is carried out on the Internet. Administration and management tasks that once used paper filing systems have been formalized into computerized procedures. Sophisticated integrated databases are utilized, with digitized plans linked to relevant data such as photos, matrices, context records, finds information and so on. Databases afford convenient storage as well as useful analytical tools for dealing with large amounts of data that were once stored on shelves or in cabinets. These digital repositories of data are linked to word processing programs for the writing of reports and illustration software for the creation of figures. Stratigraphic matrices are often constructed using specialist software. Archaeological illustrations are increasingly created using Illustrator, Corel Draw or Autocad. Photos are rendered and processed on Photoshop or other digital image editing programs. Educational presentations and talks are prepared and delivered with MS PowerPoint or Prezi. Perhaps the area of most rapid recent growth, however, is visual modeling of data using geographical information systems or GIS (Wheatley and Gillings 2000; Chapman 2006). For reflexive accounts of some of the implications of these changes for archaeology generally, see the various contributions in Clack and Brittain (2007), Moser and Smiles (2008) and Shanks and Webmoor (2012).

What this means in a practical sense is that most archaeologists today, when not actually working out on site, spend a large proportion of their time at the computer. In commercial units I estimate (from personal experience of working in the role) that project managers spend on average about 70–80 percent of their working time at the screen. The pace of much of what they do is greatly accelerated as a result of the wholesale adoption of digital technology. Much work is made easier and quicker, but this leads to a corresponding expectation that more will be accomplished, arguably resulting in a more pressured working environment.

Project officers and supervisors who are partly based out in the field typically start work in the early morning at their computers, printing out any maps, plans or other digitally generated data they might need for the day ahead. Then after the day's work on site they return to the computer in the evening to download photos, data from total stations and other digital material, or to transpose data on recording sheets into digital form. When it comes to writing reports, a more prolonged return from field to the computer is necessary. While it used to be the case that a specialist illustrator would produce maps and other illustrations for the report, it is now increasingly expected that all field archaeologists of supervisor level and above should be proficient in using relevant software to create their own illustrations. The political economy of doing archaeology as described by Berggren and Hodder (2003) is rapidly changing, and even small changes in technology can have radical effects. Computers have undoubtedly facilitated major changes in the rhythms and tempos of office-based work, the allocation of tasks and thus the structure of the archaeological organization itself.

To a certain extent this reflects the general transformation of the office that has taken place across the whole spectrum of professions over the last few decades. But archaeologists, far from merely keeping up with broader trends, have also been proactive in bringing about some of those very changes. The potential of computers for advancing ways that archaeological data might be represented and analyzed, or the past visualized and modeled, was recognized by archaeologists early on. They have been quick to take advantage of developments in computer technology and highly innovative in adapting advances to their own projects. Most are skilled in multiple aspects of computer use, in addition to the more traditional craft skills of archaeological practice. Computerization in archaeology is best thought of as an ongoing process, with "individuals and organisations constantly re-thinking and re-working their use of computers and specific applications" (Lock 2003, xiii), though of course computers—like all material culture—can trap workers into habitual patterns as well as liberate and empower them to work in new ways.

There are some areas of archaeological practice, however, where the impact of computer technology is not nearly so great. Consider for example the wet-lab (typically situated next to the main unit offices) where the sieving of soil samples takes place. Here occurs what might be described as controlled flows of different kinds of materials—soils poured out of sample bags into and out of a series of sieves with meshes of graded sizes, running water, seeds and charcoal scooped from the surface of flotation tanks to be dried on heaters, heavy residues of large particle size (caught in the sieves) systematically searched through, and unwanted residues flushed away or otherwise discarded. Data produced here will ultimately be stored on computer databases, but there are no computers in the wet-lab itself. This is partly because the environment is unsuitable for electronic equipment. But it is also because the wet-lab is where some basic and elemental contact between embodied investigators and flows of materials is going on. This previrtual world is messy, dirty, wet, muddy, dusty, steamy, alternately hot and cold. Unlike the very different world neatly framed by the computer screen, it is cluttered by numerous items of lab equipment, substances used in processing and materials brought in from site, and inhabited by the embodied presences of the wet-lab workers themselves.

Out on site, some aspects of archaeological fieldwork like surveying have been greatly transformed. Manually operated theodolites have largely been replaced by computerized total stations and various kinds of GPS device. Geophysical prospection makes use of digitally generated data to enable detection of buried structures and patterns, and this is increasingly used prior to or alongside excavation to give some indications as to where to dig. Recording systems have been aligned to the categories of computerized databases. But excavation itself—the core practice of the archaeological profession—has proved relatively impermeable to computer-oriented ways of seeing and doing. Despite numerous attempts to bring computers into the field for the accomplishment of recording tasks, the traditional material culture of excavation itself—that relatively basic technology of trowels, spades, mattocks, hand-shovels, buckets, nails, wheelbarrows, etc.—has stayed more or less the same. Site plans and sections are still for the most part measured and drawn by hand, and recording sheets filled in with handwritten notes and sketches. Digging remains a manual set of embodied craft practices. Not counting the presence of digital cameras, which are widely used, GPS devices, total stations and now ubiquitous mobile phones, excavation is actually one of the very few scientific workplaces where digital technology (and modes of visualization entailed in its use) is largely absent.

There are of course exceptions to the general rule. Framework Archaeology attempted to bring computers to the "trowel's edge" on the large-scale excavation at Heathrow Airport's Terminal 5, with computer operators working on-site inputting, managing and analyzing data (Beck 2000). More recently iPad tablet computers were successfully employed for recording excavated evidence at Pompeii, as shown in Figure 2.1. Director Steven Ellis states that the portability of iPads, their touchpad screen and absence of moving parts, makes them especially suitable for excavation recording. These were used for context recording, planning, construction of stratigraphic matrices and other tasks (Ellis 2011)—doing away with the need for context forms, planning paper, large planning boards, pens and paper notebooks. Much pioneering work with on-site computers and the associated dissemination of digital data on the Internet has been carried out at the international excavation of Çatalhöyük in Turkey (Hodder 2000; Tringham, forthcoming). Roman Silchester is another site where computers have played a major and innovative role in archaeological recording (Fulford et al. 2010; Warwick et al. 2009). But all these sites are exceptional, in the sense of having the benefit of high levels of sponsorship or funding to provide necessary infrastructure. In the case of Pompeii, Apple supplied the iPads, getting valuable publicity in return. Here I focus on what I take to be the more normal situation. Most excavations that I work on, at least at the time of writing, are relatively computer-free.

Given that most data end up in digital format, the largely nondigital environment and predominance of noncomputerized modes of perception out on site seem strange, in need of some explanation. It is clearly not the archaeologists themselves who are resistant to new ways of doing things, since many move readily and easily from spade-work in the field to screen-work in the office and back again on a daily basis. An obvious reason for the absence of computers on site is the weather. Computers have simply not been designed to be used in conditions of heavy rain or wind-blown dust. They cannot be operated easily by operators with dirt-encrusted fingers and dripping clothes. Ports and sockets are susceptible to the influx of mud, water or grit. Flowing

Figure 2.1 The use of iPads at Pompeii excavations, 2010. Photo reproduced by courtesy of Steven Ellis, University of Cincinnati.

materials are as abundant out on site as they are in the wet-lab. Development of more durable and weather-resistant technology, however, may encourage the greater presence of computers on site in future.

A more fundamental reason is the strong tactile dimension to fieldwork and the distinctly tangible nature of much of the evidence that emerges. It has weight, solidity, texture and consistency to it as well as mere visibility. During excavation, the sense of touch—as well as active manipulation and shaping of materials—is routinely used alongside vision to make sense of unfolding evidence. Many craft skills of archaeological interpretation are embodied competences, learned and enacted through sequences of muscular movements that are irreducible to computerized procedures. Even a simple operation like following along the cut of an archaeological feature such as a ditch or posthole involves manual skill and dexterity in turning the angle of the trowel blade and adjusting the strength of scraping actions to fit the particular characteristics of material being worked. Full appreciation of the textural and tactile characteristics of material evidence encountered—its roughness, wetness, coldness, plasticity, malleability, softness, hardness and so on—discourages use of computers, which tend to privilege vision over all the other senses.

This implies that those archaeologists who characteristically set out from the computerized office each morning to work out in the field and then return in late afternoon or evening to the office once more—from screen-work to spade-work and back to screen-work—are effectively stepping in and out of different modes of perception. I suspect that a fruitful line of ethnographic enquiry would be to follow archaeologists as they make the transition from one mode of perception to the other. There is not the space for that in this paper, however. Here I have the more limited objective of describing in outline new virtual sites for archaeological exploration and discovery that have opened up—new forms of visualization of emerging evidence that occur on-screen in digital space—and to compare these with the "acts of discovery" that still take place on-site as already documented in my 1991 ethnography.

NEW FORMS OF VISUALIZATION

Case Study 1

Observe an archaeologist at work on site scraping over an area of ground surface in search of features, and see objects and patterns emerge from under the moving blade of the trowel. She is situated within the material field she is working upon, her feet and knees resting firmly on the very surface that is being worked. In addition to the trowel being used, she is surrounded by other tools of the trade, such as spades, string, nails, mattocks, buckets and wheelbarrows, all of which are deployed from time to time in the accomplishment of the task at hand. Embodied in the scene, inhabiting the working space, she crouches low to get close to the evidence, her attention fixed on the area of ground immediately in front of her body where new forms are coming to light. Each scraping action with the trowel removes some material to reveal new patterns and objects. Soil boundaries and edges of features appear and disappear, some of which can be followed along laterally, downwards, or sideways at an angle dipping under other layers, leading her on to uncover further evidence that is presently hidden (for numerous and more detailed accounts of similar acts of discovery, see Edgeworth 1991, 2003).

Case Study 2

Now shift to a very different working environment. Stand behind an archaeologist working at a computer in the office and look over her shoulder as she searches for new sites on aerial photos and satellite images. As in the previous example, she too is embodied in the scene, but is in this case sitting on a chair in front of a computer and monitor on a desk in an office, with cables, pencils, papers, printer and other accoutrements arrayed all around. Her right hand holds the mouse and her forefinger rapidly switches to and from left-click and right-click buttons. She is face-to-face with the computer screen. It is a very different working posture to that of the archaeologist out in the field, but it enables her to focus on that area (of the screen) in front of her body where new forms and patterns are coming to light. In moving the cursor (with the mouse), she displays something of the deftness and speed that the field archaeologist does with a trowel, opening up new digital spaces at various scales with the click of the button.

[*In the particular instance that I observed for the purposes of writing this paper, the archaeologist/computer operator is searching for former artificial river channels over a broad area of about 200 square kilometers, attempting to distinguish between these and silted-up natural river channels. She is using ArcGIS and Google Earth in separate windows, on the same screen—toggling between them in order to look at differing views and representations of landscapes.*]

Whereas the archaeologist out in the field uses only the most basic technology of trowels, spades, buckets, etc., here by contrast there are many levels of technology intervening between the investigator and that which is investigated. The cables connecting the keyboard and mouse to the computer, and the computer to the monitor, are physical reminders of this. In using Google Earth, moreover, she is interacting with an online repository of aerial photos through the medium of unseen networks of fiber-optic cables and satellite links. The apparently smooth connections between the actions of the hand and movements of the cursor are facilitated by multiple calculations carried out by computer microprocessors in numerous locations.

[*As well as switching between applications at lightning speed, she also rapidly moves in and out at different scales. Spotting a feature worthy of interest on the regional scale, she zooms in at the click of a button, views it from a variety of different angles, zooms out to examine it in the context of other features, and then zooms in to greater detail again—all in a matter of split seconds. Or sometimes she moves from one location to another, again at great speed, to compare different types of watercourses or identify similarities in form.*]

The situation, however, is paradoxical. In one sense the archaeologist is separated and detached from the evidence by the borders of the screen, with the on-screen evidence rendered by the computer primarily in visual form. She has no direct contact with that evidence, which cannot be physically touched. Yet in another sense she displays all the attributes of a craft practitioner, demonstrating embodied skills of computer use alongside intellectual reasoning in the ongoing investigation into former watercourses. Despite the high level of technology intervening between her actions on the keyboard and what happens on the screen, the degree of hand-eye coordination displayed is impressive. She is in touch with the mouse and keyboard, and arguably in touch through these with the moving landscapes on-screen. Multiple feedback loops somehow combine these into a single field of action and perception. Impressive too is the degree of alertness to traces of former rivers barely perceptible to the untrained eye—the result of many months of practice—and this visual skill is inextricably bound up with the deftness of the fingers in their extraordinary dance on the keyboard and mouse in front of her, often touching the screen directly in order to point to or follow along features identified. It is clearly an embodied performance, rather than just an abstract mental one separate from the body. The investigative stance, albeit seated, is hands-on and engaged rather than detached and disengaged. She is no mere passive observer of an unfolding scene. Like archaeologists excavating a site, she is in partial control of the manner in which previously unseen evidence unfolds in that practical space in front of her.

[Having identified a particular site of interest on Google Earth she toggles to the ArcGIS window—switching different layers of data on and off according to type of information required. This enables her to overlay historic maps or LIDAR images over aerial photos, and to stretch any of these over 3-D models of terrain. Then she goes back to Google Earth to view the site at different times—using the "timeslider" function—so that the cropmark appears and disappears according to weather conditions, time of day the photo was taken, levels of flooding at the time and so on, all the while interpreting the new evidence in terms of its relative significance to her particular project, and then "drilling down"—her words, not mine—to investigate further anything considered to be of interest.]

This ability to switch from moment to moment between applications, between layers, between scales, between locations, between maps and aerial photos, between angles of view and between times the aerial photo was taken makes GIS software an extraordinary platform for the exploration of landscapes and the discovery of new archaeological sites. The "multiple windows" of screen-work (Friedberg 2006, 146) enable the user abruptly to change points of view within a virtual environment or from one virtual environment to another, or to hold several points of view at the same time. While staying seated in the chair and with her gaze remaining fixed on the screen she travels in a limited way through time as well as space, opening up spaces and closing them down at will, or virtually occupying multiple spaces at the same time. Freely available to download and use from the Internet, software programs like Google Earth and Flash Earth bring huge archives of aerial photographs and satellite images within reach of archaeologists everywhere—providing interfaces for searching through them at speed as if ranging across the surface of the planet, or flicking through them like a virtual deck of cards. It is hard to believe that only a few years ago one would have to get permission from an aerial photo library in advance, actually visit the library at a prearranged time, look up dates of aerial surveys and aircraft-runs on a chart before taking selected paper copies of individual aerial photos back to a table to examine. What used to take weeks can now be done in a matter of minutes.

[She repeats the procedure described earlier with other sites that catch her eye in ranging over the sweep of landscape—which moves while having the appearance of staying still, as though it is the observer who is flying over it at a particular altitude and speed, which can be changed at will. From my point of view, as observer of both the moving landscape depicted on the screen and the person seated on the chair who is operating the computer, whose fingers are rapidly whirling the scroll wheel on the mouse, it is difficult to keep up with the sheer profusion of what is being discovered. I watch in amazement as numerous former watercourses both natural and artificial—and myriad mixtures of both—come to light one after another, to be assessed and plotted in the short time I am watching the skilled operator at work. Some of these have never been documented or even noticed before. To all intents and purposes they can be regarded as new features or sites.]

Kelty (2008) explores the impact of free software in other fields. The impact of free software in archaeology cannot be underestimated. Since being made freely available in 2005, Google Earth has transformed the way that spatial data is visualized and greatly increased the number of archaeological sites being discovered. In the specific study being momentarily observed here, the initial discovery of an artificial or semi-artificial channel observable from a height of, say, 1 km eye altitude, might be followed by zooming in to scan along it from about 200 m up, revealing associated features alongside or within it. This information might then be used to direct archaeological investigators out in the field to the most potentially productive sites for excavation or ground survey, and results of that fieldwork would feed back into the computer study. Feedback loops that pertain between screen-work and spade-work are important, linking together what might otherwise be taken as two separate or unrelated types of activity. Crucially, excavation out in the field may result from exactly the kind of computer-aided identification of sites described here.

Remote imaging makes possible the viewing of areas not previously accessible, extending the range of vision into surprising domains—as for example in Adrian Myers's work on the Camp Delta prison camp at Guantanamo Bay in Cuba (Myers 2010). Myers used Google Earth and its historic imagery function to map the development of installations at the controversial camp, bypassing military restrictions. Google Earth has also been of great use in mapping the extent of looting of ancient sites in Libya (Contreras and Brodie 2010).

The great potential of the use of aerial photos and satellite imagery (and other remote sensing techniques such as LIDAR) is shown by recent work undertaken on Egyptian landscapes by Sarah Parcak of University of Alabama (BBC 2011). Using computer analyses of infrared photographs taken by powerful cameras from satellites orbiting at an altitude of 700 km, Parcak discovered no fewer than 17 pyramids, one thousand tombs, upwards of three thousand mud-brick houses and a large part of the street plan of the ancient city of Tanis, all buried under sand or river silts. Some sites thus identified have subsequently been found by excavation, confirming the veracity of the method. Parcak believes that there are thousands more sites buried under the silt of the Nile River, hidden in the detail of satellite imagery. Techniques of analysis developed could be applied to the study of archaeological landscapes in other countries as well, leading to further discoveries on a similar scale (Parcak 2009).

RETHINKING ARCHAEOLOGICAL DISCOVERY

I started writing this chapter in the belief that true archaeological discovery takes place only in relation to actual material evidence, not representations of that evidence. I used to think that discoveries made based on aerial photos, for example, were only pale shadows of real "acts of discovery" that take place on site (Edgeworth 1991, 2003). I would never have thought that computer representations of evidence in the form of aerial photos or satellite images could have the capacity to resist and challenge applied theories and ideas in the same way that more tangible materials encountered out on site do. Nor would I have believed that a computer operator could build up the same kind of interactive engagement with unfolding evidence as an archaeologist does with emerging materials in the process of excavation. For me the problem was partly one of authenticity (Rossi 2010): Somehow an encounter with a modeled version of reality could never in my view be as authentic as an encounter with the unmodeled reality itself. I still have reservations about certain aspects of simulated realities, especially with regard to biases and distortions built into software design, and feel the same disquiet that Turkle (2009) does with regard to the dangers inherent in treating simulations as real. However, partly as the result of observing the archaeologist at work on the computer in Case Study 2, my view has changed. It is clear that a general rethinking of archaeological discovery is necessary, taking due account of computers and the Internet as intrinsic elements of the mixture of human and nonhuman flows, forces and materials that together make up contemporary archaeological assemblages and encounters.

Digital landscapes like those on Google Earth go far beyond mere representations. There is an inexhaustible richness and depth to the data that parallels the complexity of evidence encountered out on site. Here features and sites can be hidden just as surely as in the material landscape, brought to light only through the performance of the craft skills of archaeology—though with the mouse and cursor taking the place of the trowel and spade.

If probed with skill and discernment, the virtual landscape can potentially yield an almost infinite number of new discoveries, each one giving rise to further paths of exploration that can be followed toward further discoveries and insights.

The example of the computer operator discovering new sites in an on-screen landscape poses ethnographic conundrums as well as archaeological ones. As Marc Augé puts it, "all ethnology presupposes the existence of a direct witness to a present actuality . . . Anything remote from direct observation . . . is not anthropology" (Augé 2008, 8). Yet in Case Study 2 both the ethnographer and the ethnographic subject are direct witnesses not to a present actuality but to a "virtual actuality" on-screen. The landscape encountered and explored is not the "actual" landscape, at least not in Augé's sense of the word. It is a highly shaped and prepackaged version of the landscape, made up from aerial photos that are stitched together by Google Earth using digital terrain data collected by NASA's Shuttle Radar Topography mission, all processed electronically and channeled through fiber-optic cables to the computer. That of itself is not a problem: It has long been realized by STS practitioners that most scientific discovery takes place on materials that have been in some way prepared beforehand, and that are full of the artifacts of the preparation process (Lynch 1985). Thin-sectioned slices of muscle tissue placed on a microscope slide and viewed through powerful optical lenses are in some ways just as artificial as moving images on a computer screen are. The problem is more the practical question of where to situate the ethnographic study itself.

Should one situate the ethnography so that the object of study consists of the interactions between the subjects and the apparatus they are using? That is one prosaic solution, though not an entirely satisfactory one, precisely because the significant interactions are taking place on-screen, beyond the material domain of the hardware that makes them possible. But does this mean one should therefore shift the locus of ethnographic study onto those virtual interactions, and how could this realistically be achieved? During one early ethnographic study of computer scientists, investigators noted of computer scientists that "their information gathering and communications appear largely to be carried out in the virtual world provided by the Internet" and posed the question "How can I enter this world and yet retain the ethnographer's scepticism towards the taken-for-granted features of this technology?" (Cooper et al. 1995, 20).

To turn the question around, we might ask whether the ethnographer has to be a "subject"—an "I"—at all. Some of the roles of the ethnographer can clearly be delegated to computers, in recording complex sequences of keyboard actions or plotting eye movements across the screen, for example. Likewise, we might ask whether the ethnographic activity that is being studied has to always include a human being. Object-object interactions are taking place alongside and in some cases replacing human-object interactions (for further discussion on some of these issues, see Kelty 2009).

Many assumptions about participant-observation are challenged by use of computers. Can the human ethnographer achieve the kind of cultural immersion in virtual space necessary to write thick descriptions of the interactions that take place there, without losing his or her grounding in the embodied reality of here and now? Indeed, is it actually possible for an ethnographic study to be carried out on virtual and actual levels of reality simultaneously? That would be roughly equivalent (in the film *The Matrix*, when Neo is offered the choice by Morpheus) to choosing to take both the red pill and the blue pill at the same time. Perhaps the answer is to attempt to embrace both realities by shifting back and forth from one to the other— to toggle, so to speak, between one's embodied experience of the concrete, tangible office-setting and the disembodied interactions with immaterial, intangible landscapes that move across the computer screen.

Participation in the virtual interactions taking place on-screen would seem to be a necessary element of any in-depth ethnography of archaeologists at work on computers. Julia Gillen has written an innovative short ethnographic account of virtual archaeology on Second Life, as part of a study of visual literacy among teenagers. She describes her interactions with teenagers she had never met face-to-face, and their joint adventures in discovering artifacts like wooden chests from shipwrecks on the sea-floor (through the medium of personal avatars in a simulated 3-D environment). She emphasizes that making sense of a database of chat-logs would be much more difficult if she had not immersed herself and participated in unfolding Second Life scenarios, though she acknowledges the extra experiential load this placed on her role as researcher (Gillen 2011). Interestingly, Gillen draws from and makes connections with ethnographies of actual archaeological practices (Goodwin 2006; Van Reybrouck and Jacobs 2006) in formulating her ethnographic method for dealing with discoveries and interactions in digital space. This illustrates the extent to which—in discussing the assimilation of computers into archaeological assemblages— there is also a movement in the other direction. Assimilation works both ways. Archaeological tools, methods, perspectives, sites and artifacts are themselves being incorporated into digital and virtual assemblages, transposed into other domains, influencing computer visualization in numerous subjects and areas of experience outside of academia.

Both archaeological and anthropological ontologies have to be rethought in the light of the widespread influence of computers and digital communication networks. A basic and obvious difference between the archaeologist excavating on site (Case Study 1) and the archaeologist searching for new sites on-screen (Case Study 2) is that the latter is physically linked to a global digital network connecting countless microprocessors around the world, not to mention the hundreds of thousands of miles of underground cables, and scores of satellites spinning through orbital space—whereas the former is not. One worker is plugged in and gridded in, so to speak, while the other is unplugged and ungridded (though she is of course part of other networks and grids). The unplugged archaeologist is attending to the immediate material environment within which she herself is embodied and situated, using the sense of touch alongside vision. By way of contrast, the plugged-in archaeologist is engaging and interacting with a landscape totally removed from her own embodied situation in the here and now. Although she is using embodied skills and multiple senses in physically engaging with the computer hardware, the displacement of archaeological reality onto the screen prioritizes vision and excludes all but the shadows of other forms of sensory experience (though future advances in virtual reality may redress that sensory imbalance).

There is much that is different but, perhaps surprisingly, there are significant similarities too. Both on-screen and on-site work constitute places of discovery, in the sense that something real and insistent from outside realms of knowledge is pushing through, emerging, taking shape—manifesting itself in the context of practical work upon it. Ideas are applied in a kind of physical and cognitive wrestle with unfolding objects, patterns, sites. Investigators in both cases are using embodied craft skills to open up spaces within which they engage and interact with emerging evidence, asking questions of it and probing into it, whether with cursor or trowel. As a result of this probing, things that were previously unseen—hidden in the earth or buried in the incredible detail of aerial photos and satellite imagery—are coming to light. Each of the different kinds of encounters between the known and unknown can be characterized as mixtures of the human and nonhuman, or as changing assemblages of the social, material and technical. In moving from spade-work to screen-work, the archaeological assemblage is reassembled, with new elements and components added to the mix and others taken away, but the structure of the act of discovery essentially remains intact, even when transposed from direct embodied experience into virtual space. For discussion on how archaeological reality and cyberspace—treated largely as separate "sites" in this paper—are fast becoming integrated hybrid realities, see Forte (2010).

Further transformations of human-computer relations in archaeological assemblages are happening now and will continue to occur in the future. As I write this chapter, I am reading of recent work on Northern Mesopotamia that, like Parcak's research outlined earlier, makes use of the incredible detail on satellite images of the Earth's surface. Archaeologists from Harvard together with computer specialists from MIT have discovered in those images over nine thousand new sites within an overall area of about 23,000 square kilometers. Sites are mostly close to rivers and range from Neolithic tells (settlement mounds made up of the gradual accumulation of occupation debris) to multiperiod "kites" (stone walls crossing the desert that intercepted migratory flows of animals and funneled them into enclosed killing zones). The authors claim to have the capability to make a comprehensive map of most archaeological sites within an area so vast that it would take teams of archaeological surveyors working on the ground many years to cover. What makes this research especially interesting is that computers are being used for more than mere modeling and analysis of data, or for facilitating discovery. Software applications have been devised for scanning and searching of images for discolorations in the soil that indicate the presence of "anthrosols" (humanly modified soils) and associated settlement sites. In other words, *part of the work of archaeological discovery has been delegated to the computer.* This greatly speeds up the process of discovery and makes possible identification of sites over very large surface areas, arguably doing away with the need for large-scale surveys undertaken from the ground (Menze and Ur 2012). The act of discovery, it seems, at least when defined as the initial encounter with emerging evidence, does not necessarily have to be carried out by human beings. Computers in this instance have been programmed to do it for them.

CONCLUSION

This main question asked by the chapter might be framed as "How are our notions of discovery changed when computers and associated networks are taken into account as intrinsic parts of archaeological assemblages?" As already noted "assemblages" are understood here to include tools used in the production and reproduction of knowledge, as well as ancient artifacts and features found in the ground. The ethnography of an excavation I carried out in 1991 showed how our understanding of the past was inseparable from the tools used to unearth it, but that was taking account only of things like trowels, spades and planning grids. Now it is necessary to expand that rather limited ethnographic view of archaeological assemblages to include computers and computerized networks too—and thus to rethink the theory of archaeological discovery accordingly. The ontological ground of a discipline like archaeology shifts as the specific mix of human and nonhuman in its assemblages is reconfigured: We should perhaps think of our understanding of it (rather like the process of computerization itself) not as something fixed and static but as a work in progress, never taking a finished form but always in the process of being updated and renewed to take account of new developments.

The fact that computers are now important elements of archaeological assemblages has effects far beyond spheres of discovery. The extraordinary power of computers for modeling and visualizing archaeological data (only some aspects of which have been touched upon here) leads inevitably to shifts in organizational and political structures of the discipline. Funding that used to go to excavations increasingly gets diverted to computer visualization. Postgraduate courses in computer techniques in archaeology, especially GIS, may be better subscribed than those that cover more traditional aspects of archaeological practice. Many archaeologists with doctorates have great experience of screen-work yet practically none of spade-work. Indeed, an increasing number of archaeologists argue that surface survey (aided by computers, leading on to greater use of computer visualization) now provides an alternative trope for thinking about contemporary archaeology. The traditional view of "archaeology-as-excavation," with its implicit ontology of emergence and discovery, not to mention its grounding in multisensory reality, is replaced by an alternative perspective on "archaeology-as-surface-survey" (Harrison 2011). Archaeological practices in these terms are regarded as processes of assembling and reassembling—somewhat in the manner that virtual landscapes are assembled and reassembled on-screen—rather than delving into the ground to make buried things come to light (for a counterview, see Edgeworth 2011).

It would be wrong, however, to see screen-work and spade-work as somehow opposed to each other, with separate ontologies, when they are linked activities that intermesh as different parts of the same general archaeological enterprise. Understandings of archaeological ontology should perhaps encompass both. Discoveries on-screen direct archaeologists to sites with the most potential for excavation or detailed ground survey out in the field. Results of excavation in turn provide crucial feedback and further impetus for on-screen-work. Computers and global networks of communication are increasingly embedded, alongside spades and wheelbarrows and more ancient material culture, as everyday constituents of the material assemblage of contemporary archaeology, rapidly subsumed into and transforming the ontological structure of discovery itself.

REFERENCES

Augé, Marc. 2008. *Non-places: An Introduction to Supermodernity.* 2nd English ed. Translated by John Howe. London: Verso.

BBC. 2011. "Egyptian Pyramids Found by Infra-Red Satellite Images." Accessed March 15, 2012. http://www.bbc.co.uk/news/world-13522957. Last accessed 4 November 2013.

Beck, Anthony R. 2000. "Intellectual Excavation and Dynamic Information Management Systems." In *On the Theory and Practice of Archaeological Computing,* edited by Gary Lock and Kayt Brown, 73–88. Oxford: Oxbow.

Bennett, Jane. 2010. *Vibrant Matter: A Political Ecology of Things.* Durham, NC: Duke University Press.

Berggren, Åsa, and Ian Hodder. 2003. "Social Practice, Method, and Some Problems of Field Archaeology." *American Antiquity* 68 (3): 421–434.

Chapman, Henry. 2006. *Landscape Archaeology and GIS.* Stroud: Tempus.

Clack, Timothy, and Marcus Brittain, eds. 2007. *Archaeology and the Media.* Walnut Creek: Left Coast Press.

Contreras, Daniel A., and Neil Brodie, 2010. "Shining Light on Looting: Using Google Earth to Quantify Damage and Raise Public Awareness" *SAA Archaeological Record* 10 (3): 30–33.

Cooper, Geoff, Christine Hine, Janet Rachel and Steve Woolgar. 1995. "Ethnography and Human-Computer Interaction." In *The Social and Interactional Dimensions of Human-Computer Interfaces,* edited by Peter J. Thomas, 1–36. Cambridge: Cambridge University Press.

Cowgill, George L. 1967. "Computer Applications in Archaeology." *AFIPS Conference Proceedings* 31: 331–338.

Deleuze, Gilles, and Felix Guattari. 1987. *A Thousand Plateaus*. Translated by Brian Massumi. Minneapolis: University of Minnesota Press.

Edgeworth, Matt. 1991. "The Act of Discovery: An Ethnography of the Subject-Object Relation in Archaeological Practice." Doctoral thesis, University of Durham. Accessed March 18, 2012. http://etheses.dur.ac.uk/1481.

Edgeworth, Matt. 2003. *Acts of Discovery: An Ethnography of Archaeological Practice*. BAR International Series 1131. Oxford: Archaeopress.

Edgeworth, Matt. 2011. "Excavation as a Ground for Archaeological Knowledge." *Archaeological Dialogues* 18 (1): 44–46.

Ellis, Steve. 2011. "iPads at Pompeii." Porta Stabia. Accessed April 15, 2012. http://classics.uc.edu/pompeii/index.php/news/1-latest/142-ipads2010.html.

Falzon, Mark-Anthony, ed. 2009. *Multi-sited Ethnography: Theory, Praxis and Locality in Contemporary Research*. Farnham: Ashgate.

Forte, Maurizio. 2010. *Cyber-archaeology*. BAR International Series 2177. Oxford: Archaeopress.

Friedberg, Anne. 2006. *The Virtual Window: From Alberti to Microsoft*. Cambridge, MA: MIT Press.

Fulford, Michael, Emma E. O'Riordan, Amanda Clarke and Mike Rains. 2010. "Silchester Roman Town: Developing Virtual Research Practice 1997–2008." In *Digital Research in the Study of Classical Antiquity*, edited by Gabriel Bodard and Simon Mahony, 15–34. Farnham: Ashgate.

Gillen, Julia. 2011. "Rethinking Literacies, Learning and Research Methodology around Archaeology in a Virtual World." Accessed April 14, 2012. http://eprints.lancs.ac.uk/52114/1/Gillen_final.pdf.

Goodwin, Charles. 2006. "A Linguistic Anthropologist's Interest in Archaeological Practice." In *Ethnographies of Archaeological Practice: Cultural Encounters, Material Transformations*, edited by Matt Edgeworth, 45–55. Lanham: Altamira.

Gupta, Akhil, and James Ferguson. 1997. *Anthropological Locations: Boundaries and Grounds of a Field Science*. Berkeley: University of California Press.

Harrison, Rodney. 2011. "Surface Assemblages: Towards an Archaeology in and of the Present." *Archaeological Dialogues* 18 (2): 141–196.

Hodder, Ian, ed. 2000. *Towards Reflexive Method in Archaeology: The Example at Çatalhöyük*. Cambridge: McDonald Institute for Archaeological Research.

Kelty, Christopher M. 2008. *Two Bits: The Cultural Significance of Free Software*. Durham, NC: Duke University Press.

Kelty, Christopher M. 2009. "Collaboration, Coordination, and Composition: Fieldwork after the Internet." In *Fieldwork Isn't What It Used to Be*, edited by James D. Faubion and George E. Marcus, 184–206. Ithaca: Cornell University Press.

Latour, Bruno. 1993. *We Have Never Been Modern*. Cambridge, MA: Harvard University Press.

Lock, Gary R. 2003. *Using Computers in Archaeology: Towards Virtual Pasts*. London: Routledge.

Lock, Gary R., and Kayt Brown, eds. 2000. *On the Theory and Practice of Archaeological Computing*. Monograph 51. Oxford: Oxford University Committee for Archaeology.

Lynch, Michael. 1985. *Art and Artifact in Laboratory Science: A Study of Shop Work and Shop Talk in a Research Laboratory*. London: Routledge.

Marcus, George E. 1995. "Ethnography in/of the World System: The Emergence of Multi-sited Ethnography." *Annual Review of Anthropology* 24: 95–117.

Menze, Bjoern, and Jason Ur. 2012. "Mapping Patterns of Long-Term Settlement in Northern Mesopotamia at a Large Scale." *Proceedings of the National Academy of Sciences of the United States of America* 109: 12.

Moser, Stephanie, and Sam Smiles, eds. 2008. *Envisioning the Past: Archaeology and the Image.* Oxford: Blackwell.

Myers, Adrian. 2010. "Camp Delta, Google Earth and the Ethics of Remote Sensing in Archaeology." *World Archaeology* 42 (3): 455–467.

Parcak, Sarah H. 2009. *Satellite Remote Sensing for Archaeology.* London: Routledge.

Renfrew, Colin, and Paul Bahn. 2008. *Archaeology: Theories, Methods, and Practice.* 5th ed. London: Thames and Hudson.

Rossi, Michael. 2010. "Fabricating Authenticity: Modeling a Whale at the American Museum of Natural History, 1906–1974." *Isis* 101 (2): 338–361.

Shanks, Michael, and Timothy Webmoor. 2012. "A Political Economy of Visual Media in Archaeology." In *Re-presenting the Past: Archaeology through Image and Text*, edited by Sheila Bonde and Stephen Houston, 87–110. Oxford: Oxbow Books.

Tringham, Ruth. Forthcoming. "Forgetting and Remembering the Digital Experience and Digital Data." In *Excavating Memories*, edited by Dusan Boric. Oxford: Oxbow Books.

Turkle, Sherry. 2009. *Simulation and Its Discontents.* Cambridge, MA: MIT Press.

Van Reybrouck, Dirk, and David Jacobs. 2006. "The Mutual Constitution of Natural and Social Identities during Archaeological Fieldwork." In *Ethnographies of Archaeological Practice: Cultural Encounters, Material Transformations*, edited by Matt Edgeworth, 33–44. Lanham: Altamira.

Warwick, Claire, Melissa Terras, Claire Fisher, Mark Baker, Emma O'Riordan, Matt Grove, Mike Fulford, Amanda Clarke and Mike Rains. 2009. "iTrench: A Study of User Reactions to the Use of Information Technology in Field Archaeology." *Literary and Linguistic Computing* 24: 211–223.

Webmoor, Timothy. 2005. "Mediational Techniques and Conceptual Frameworks in Archaeology: A Model in Mapwork at Teotihuacan, Mexico." *Journal of Social Archaeology* 5 (1): 54–86.

Wheatley, David, and Mark Gillings. 2000. "Visual Perception and GIS: Developing Enriched Approaches to the Study of Archaeological Visibility." In *Beyond the Map: Archaeology and Spatial Technologies*, edited by Gary R. Lock, 1–27. Amsterdam: IOS Press.

Wittel, Andreas. 2000. "Ethnography on the Move: From Field to Net to Internet." *Forum: Qualitative Social Research* 1: 1. Article 21. Accessed March 15, 2012. http://www.qualitative-research.net/index.php/fqs/article/view/1131. Last accessed 4 November 2013.

3 British Columbia Mapped

Geology, Indigeneity and Land in the Age of Digital Cartography

Tom Schilling

What makes a mapmaker in the age of computer-assisted visualization?[1] What new professional commitments are shaping the political subjectivities of these knowledge-workers, and how are these commitments shaping the ideas and attitudes of the people living in the regions being mapped? This chapter will explore recent attempts by two communities to produce, standardize and disseminate large-scale digital maps of contested areas throughout British Columbia, Canada. In both cases, the maps involved are compiled with data collected through a historically shifting range of methods. Once completed, they enter the public domain with explicit imperatives: Their reception can help shape the course of economic development, or validate claims to political sovereignty made by marginalized groups. For exploration geologists seeking new mineral prospects in the Canadian Rocky Mountains, computer-simulated, regional-scale models produced by Geoscience BC simultaneously help to refine and refocus the search for target sites. As abstract representations with precisely defined geographic coordinates, geological models also flatten fraught political geographies, reducing all measured territory to mineralogical values, and often delaying direct engagement with the groups who inhabit the land above would-be mines. In contrast, among many of the 203 Aboriginal First Nations bands living within British Columbia, digital maps denoting traditional place-names and knowledge have been produced to amplify long-standing territorial claims with social, historical and cultural associations, creating self-consciously essentialized composites that can act as key evidence in treaty settlement proceedings.[2] Like the maps and models produced by Geoscience BC, these maps help First Nations groups to project historical identities, future ambitions and economic designs onto visual representations of the province.

For many users, however, these maps and their accompanying data sets are merely starting points for the construction of still more maps, steps in a recursive process of representation and legitimation that often blurs the origins of databases and increasingly removes contemporary cartographers from the work of empirical observation. In this chapter, I argue that it is precisely this manipulability that distinguishes digital cartography from its paper-based predecessors: While both invariably reflect the interests and epistemological biases of their producers, computerized methods of assembly and compilation further obscure the regimenting effects of these origins by inviting new users to experiment and play. Earlier academic treatments of computer-produced, network-hosted data-images have often been quick to celebrate this intuitive accessibility of synoptic visualizations, championing the democratizing power of the Internet. By doing so, these scholars have often neglected the politics of data collection and federated standards, highlighting the agency of the individual user while ignoring the framings and elisions effected by the institutions in charge of maintaining and coordinating the data sets involved.

While the uses and meanings of geophysical models and First Nations place-names maps diverge in many ways, the frequency with which they now appear together in emergent land claims negotiations throughout British Columbia warrants consideration. In this chapter, I will review the historical transitions and institutional circumstances through which both kinds of mapmaking came to be recognized by geologists, stockholders, land developers and First Nations citizens as legitimate ways to represent knowledge and, like paper maps, how digital maps came to be used as vehicles of authority and expertise, and as proxies for otherwise inaccessible objects. The opening sections follow the political lives of geological maps in the province from the founding of the British Columbia Geological Survey in the late nineteenth century through the current era of web-based claim staking and speculative investment. The final third of the chapter is more exploratory, reviewing anthropological debates over the political expediency and broader consequences of recent attempts by First Nations groups to consolidate cultural knowledge into map-ready formalisms. In the conclusion, I outline future ethnographic work at the sites of the growing digital data archives currently being organized and promoted by several key nongovernmental organizations involved in mineralogical exploration and First Nations advocacy. Small exploration companies and land claims negotiators are challenging distinctions between "producers" and passive "users" of maps by building their own maps and constructing their own plans from these archives. Nevertheless, the increasing computerization of cartography in British Columbia has yet to fulfill the ideals of open and collaborative resource management promised by the archives' creators.

VISUALIZING CLAIMS

Vancouver serves as corporate headquarters to over 1,200 globally active mineral exploration companies (Anon. 2011c). Few public databases of fine-scale information existed, however, prior to the establishment of Geoscience BC in 2005. The organization received its initial funding from the provincial government of British Columbia, and has supported itself since then through a combination of public and private grants. Its creation was first advocated by the Association for Mineral Exploration, British Columbia (AME-BC), whose geologist affiliates were frustrated by a steady decline in federal funding for exploration-oriented geology and geophysics in southern Canada during the 1990s, a trend they attributed to the federal government's mounting interest in developing (and thus maintaining jurisdiction over) mineral claims in the country's northern territories. The geologists who founded the organization were not only scientists but also entrepreneurs who were worried by what they saw as a dearth of investment in mining and exploration throughout the province. In their eyes, fault for the lack of new developments lay with the paucity of publicly available geoscientific information, a problem that was magnified by the fact that most of the exploration companies working throughout British Columbia were private businesses with few incentives to make their raw data available for public inspection (Anon. 2011a).

Geoscience BC was not the only group to turn data access into a rallying cry at the end of the twentieth century. For both exploration geologists and First Nations advocates, the growing ubiquity of digital mapmaking tools throughout the 1990s presented new means for sharing information vital to the prosecution of business and politics. For mineral developers working in contested regions, selling stock in exploration projects, navigating environmental regulations and acquiring a "social license to operate" from local residents and broader publics soon came to mean promoting plausible futures in a number of discursive venues (Howard-Grenville, Nash and Coglianese 2008). For Geoscience BC, promoting mineral development meant collecting geological data from exploration companies and sharing them online—albeit in carefully simplified form.

Similarly, professional cartographers, hobbyists, activists and First Nations representatives who began sharing tips and techniques through the online Aboriginal Mapping Network (AMN) in 1998 learned how to train each other in the evolving conventions of traditional knowledge databasing and place-names mapping. Despite this ethic of collaboration, however, many AMN participants restrict access to the most detailed versions of their final products to the members of the First Nations whose knowledge the maps represent, generating simplified versions for broader circulation. In 1997, the Delgamuukw decision reasserted First Nations' rights to make territorial claims in British Columbia, rights that the federal government originally argued had been extinguished during the colonial period (Bankes 1998, 317). A year later, the Nisga'a nation signed a treaty accepting a land settlement of 2,000 square kilometers, and the ensuing rush of land claims lawsuits initiated by other First Nations quickly produced an entire industry devoted to claims-related data collection and map production.

For groups like the Gitxsan, Dane-zaa Doig River and dozens of other First Nations currently negotiating treaty settlements with the province, "data collection" has come to mean hiring geographic information systems

(GIS) specialists and ethnographic consultants to record oral histories, place-names and folklore from traditional cosmologies (Chambers et al. 2004; Chapin, Lamb and Threlkeld 2005). It also entails fixing these "data" to the gridded coordinate systems and satellite photos housed in the map archives of Natural Resources Canada (NRCan), the federal ministry of mapping, minerals and the environment. Once the maps are complete, their publication requires GIS specialists to formalize elaborate, often narrative-based descriptions into categories of indexical markers for the convenience of other mapmakers, who may then treat this set of markers as simply another data layer as they construct their own maps of the province. These tools have helped First Nations negotiators earn preliminary positions from which to advocate for new terms of recognition and expanded authority within environmental and economic comanagement projects. But as these data sets are taken up by other institutions and distributed through different archives, the meaning and origins of the knowledge they imply are growing increasingly obscure.

Geoscience BC, AMN and other consulting organizations currently producing maps within British Columbia do not adhere to a single corporate model. Some, like Geoscience BC, receive public funds and conduct their work with little political interference; others, like Ecotrust Canada, host of AMN, are primarily supported by private donations, and use their resources to generate information which other groups then use to lobby for environmental conservation measures and against development proposals in various regional, national and international venues. At both organizations, however, mapmakers' attempts to equate the online accessibility of their maps and data sets to public interest and democratic participation obscure the mechanics and motivations behind earlier formatting, curating and editing choices. Focusing on the institutional contexts of digital cartography and data assembly in emergent and interrelated controversies over development proposals, environmental conservation and indigenous territorial rights reveals the central importance of standardization and institutional control in the making of maps, even—especially—when the objects produced are intended for subsequent analysis and reassembly. The origins of these "black boxing" maneuvers, however, lie in the historical development of the British Columbia land claims registration process.

CARTOGRAPHY, DEVELOPMENT AND COMPUTERIZATION

Since the beginning of colonial settlement in the province in the early nineteenth century, the forbidding topography of British Columbia has played host to a series of parallel and often overlapping genres of land claims. Beginning with the first provincial gold rush on the coastal Queen Charlotte Islands in 1850, small-scale prospectors, and later exploration geologists and mining engineers, have fought to stake claims and develop mineral prospects on publicly owned land (Barman [1991] 2007, 55–76). As the Canadian government sought to solidify territorial claims on this western frontier throughout the middle of the nineteenth century, documents produced by geologists employed by the Geological Survey of Canada (GSC) began to circulate among ever-wider audiences, and the information these documents contained held increasing influence on eastern Canadians' understanding of the country's western landscapes and the monetary worth of their potential mineral resources (Zaslow 1975; VanDine, Nasmith and Ripley 1993). The maps, sketches and other visual representations of the region's landscape produced by mining engineers, geologists and surveyors throughout the twentieth century played a key role in fixing landmarks to spatial coordinates throughout the province and in representing the area as a whole as predictable: promising for investment, and "safe" for white settlement (Braun 2000; Grek-Martin 2011).

While paper-based geological maps still serve important functions and circulate widely among mining interests in British Columbia, the late twentieth-century development of computerized geoinformatics has come to provide a common visual language for spatial and quantitative aspects of developmental planning, mineral speculation and land tenure disputes. Database access, federated aesthetic standards and the bounded possibilities of improvisation and reconstruction enabled by web-based production, editing and distribution all present new questions for the critical study of contemporary mapmaking. Problems related to spatial organization, map-inflected ontology and the politics of cartographic representation have long occupied anthropologists and historians (Gell 1985; Bourdieu 1990; Orlove 1991; Harley 1992). Historians of science, however, have largely confined their treatments of cartography to the mechanical tools, imperial governments and paper-based media of pre-twentieth-century mapmaking, and the seventeenth- to nineteenth-century transition to "modern" notions of territory (Winichakul 1994; Edney 1997; Burnett 2001)—what William Rankin refers to an "international system of perfectly bounded, politically autonomous, and internally consolidated sovereign states" (Rankin 2011, 7). Contemporary mapping technologies like the Global Positioning System (GPS) rely on the stability of these earlier constructions, but by allowing diverse groups of users to assign coordinates to all manner of material and semiotic objects, they also support the creation of digital maps, models and other visualization technologies that provide detailed representations of features within and across contested domains. Some of these spatial models, such as those used within the mineral exploration industry, succeed in playing dumb to their implications in larger political conflicts about boundaries and ownership within these regions. As Rankin explains, "Boundary-crossing geographic systems create a common datum for locating and stabilizing international boundaries, but they also make it much easier to ignore these boundaries when . . . planning massive engineering works" (Rankin, 5). As a meeting ground for the geological sciences, remote sensing, satellite photography and GIS data, geoinformatics only further reifies problematic geopolitical constructions, burying them underneath mountains of tabularized, precise and seemingly unimpeachable data.

Thanks to statistical analysis, satellite photography and synoptic, Internet-based navigation tools, the practice of visually representing landscapes and geological features has become, in terms of the sheer bulk of information provided by these technologies, far more spatially determined than when knowledge of remote landscapes was captured only through sketches, descriptive narratives and atlases, and hand-drawn maps (Helmreich 2011). Some activist- and community-produced maps produced to document and protest natural gas developments, oil spills, flawed epidemiological studies in urban areas and other controversial projects and events have drawn upon these resources in novel and powerful ways (Corburn 2005; Dosemagen, Warren and Wylie 2011). The deployment of maps and other community-assembled data sets as visual arguments in public health and environmental debates has helped some groups to generate political momentum on subjects otherwise dominated by technical experts. Indigenous groups in Latin America and elsewhere have also drawn upon these techniques in "counter-mapping" projects designed to undermine the authority of developers and state surveyors (Peluso 1995; Wainwright and Bryan 2009).

Practical successes notwithstanding, critical analyses of mapmaking projects have often simplified the multifarious political futures of digitized data sets by focusing primarily on antagonisms between marginalized groups and nation-states, ignoring the mediating roles played by NGO mapmakers and data collectors and the standardizations, reformatting conventions and distribution networks they enable. In linking digital data collections to statistical models and simulations, however, some philosophers and historians have called attention to the institutional and rhetorical framings used to justify the authority, validity and comprehensiveness of computer-assisted representations and arguments, particularly when these models are incorporated into environmental management policies (Oreskes, Shrader-Frechette and Belitz 1994; Winsberg 1999; Oreskes 2007). Making sense of the professional ethics of digital cartography means attending to the material cultures that dictate the ways new communities of practitioners have learned to collect, classify and display different categories of information. It also means paying attention to the historical conditions of possibility that first lent these maps and models conceptual meaning and political agency.

PRIVATE DATA, PUBLIC MAPS

The geologists who first organized the British Columbia Geological Survey (BCGS) in 1895 recognized the risk of losing the geological knowledge accumulated by prospectors to proprietary secrecy. According to the regulations for certifying staked claims with the provincial government, prospectors must agree to provide to the BCGS a portion of the information they obtain during the exploration process in order to keep their claims active (Claims Registry 2012). In order to protect the prospector's investment, the BCGS cannot publicly disclose this received information for a brief period, and prospectors may petition the survey to withhold particularly valuable information for up to three years. Even when this protected period ends, however, the data that are then published are rarely the full extent of the information gathered at a given site. During periods of inactivity, claim-holders are allowed to pay a small fee in lieu of providing new data. As competition between developers and exploration companies increased in the second half of the twentieth century, many prospectors began paying these claim fees many years in advance in order to keep their data out of the BCGS's annual reports (Anon. 2011b). Since its inception, then, BCGS officials have been confronted by a multiplicity of maps: the large-scale surveys collected and compiled with government workers and public funds, and a vast, disconnected assemblage of unstandardized, semiprivate data sets, few of which ever entered circulation outside their parent companies, and then often only in the most hopeful and selective of forms.

Geoscience BC was originally tasked with gathering and organizing geo-chemical data from its own field expeditions, supplementing and building from maps produced through provincial and federal surveys to construct regional-scale maps of mineralogical density for a series of large areas throughout the province where the depth to bedrock, or the thickness of the glacial sediment layer, remained unknown. Accurately determining this depth was a critical step for presenting the province's minerals as accessible, and therefore financially viable: Whereas some regions remained buried beneath layers of debris hundreds of meters thick, in other areas, valuable bedrock could be reached by "scratching the surface with a backhoe" (Anon. 2011a). Nearly all of the exploration companies active in British Columbia begin their work with the regional-scale magnetic and gravity data maps composed and published by academic geology departments and federal agencies like the Geological Survey Canada (GSC), which hosts data obtained from hundreds of scans "collected from 1947 to the present" (Mira Geoscience 2011, 9) in their Geophysical Data Repository. Once exploration officials select a "target site" and commission the development of a local model, however, these larger maps are rendered innocuous, displayed as aesthetic treats on company websites and in the glossy pages of annual reports. Compiling regional-scale maps that initiate these searches is a different matter: Whereas a company attempting to plot the edges of an ore body underneath a surface area a few square kilometers in size can afford to fly instrument scans over its area at 50-meter intervals, several of the regions surveyed by Geoscience BC span tens of thousands of square kilometers.

PADDING CELLS AND SHAKING HANDS

While computer-driven data-reduction methods allow the organization's geostatisticians to deal separately with the disparate and overlapping meshes generated from the superposition of numerous public and private databases, Geoscience BC's conventions of representation governing its regional-scale models dictate that these grids must be stitched together into a single, uniform whole before the penultimate step of numerical inversion can take place. Upon the publication of the completed model in the organization's open-access online database, the obscured authorship of the composite image is represented as the diligent analytical work of a single geostatistical consultancy. As Geoscience BC expands from its original studies of lightly prospected areas into more heavily explored regions, however, the databases from which the organization can source aerial scan information are multiplying. Already well-practiced at adopting data from public repositories to match the parameters of its own coarse scans, the organization has lately begun soliciting privately collected data covering the "Golden Triangle" mining region in the northwest corner of the province. Depicting these data sets in their received forms yields an uneven patchwork of heterogeneous grids, each with different dimensions, scan line spacing intervals and geographic orientations.

For the geophysicists and instrument-towing helicopter pilots employed and contracted by the organization to collect new data, selecting an appropriate scan interval is both a scientific compromise and an economic gamble. The mineral density models that eventually arise from the data obtained from coarse scans are even more constrained by formal parameters, from the algorithms used to "smooth" data and estimate values between discontinuous points to the dimensions of the volumetric grids used to simplify the work of calculation and data reduction. If calibrated incorrectly, these grids can be fundamentally misleading, inciting drilling crews to conduct detailed studies of new areas based on vague suggestions of mineralization. To ameliorate these risks, geostatisticians spend considerable time "winnowing suspect data" and devising computational methods to join coarse data from scans conducted at 1- to 4-kilometer intervals with refined scans in select areas. This statistical scaling procedure is similar to what Eric Winsberg (2010, 72–92) refers to as the "handshake method" of combining disparate computer simulations into a single model; Timothy Webmoor (2005) draws similar parallels in his discussion of the integration of digital graphics and conventional photographs increasingly involved in archaeological "mapwork." In the QUEST-South mineral density model published online in late 2011, for instance, data obtained in south-central British Columbia were first separated into regions of regular grids discretized at 500-meter intervals, then surrounded by boundary regions of substantially coarser "padding cells" designed to generate sufficient "volume" in unexamined regions to completely fill the space covered by the map.

MODELS OF MINERALS, MODELS FOR WORK

Conscious of their roles as cartographers and visual communicators, Geoscience BC's statisticians recognize that their "models will more easily facilitate geologic interpretation and definition of favorable geology than the data alone," and that they have the added benefit of being useful "in a quantitative manner using 3D-GIS analysis" (Mira Geoscience 2011, 3–4). Hosted as tabularized data files and formatted for a wide range of modeling and analysis programs, these maps-made-whole can be downloaded and divided anew, their standardized parts redacted, refined through new additions or subject to more daring algorithms for quantitative analysis. Whereas smaller exploration companies or even individual prospectors might begin planning more detailed fieldwork based on insights gleaned through their inspections of the published models themselves, larger companies with in-house geostatistics experts often use only the numerical data published alongside the models, producing new visual models and quantitative simulations for their own private inspection. By acting as a homogenizing gatekeeper between data-hungry exploration geologists and the proliferating standards and inconsistent methods behind the digitized (and increasingly, merely digital) archives of British Columbia's disconnected mass of geological knowledge, Geoscience BC puts a premium on convenience, delivering a broad-strokes geological picture of the province to potential developers, many of whom might not have bothered to consider exploring the territory if they had had to gather these initial maps themselves.

Much like the traditional knowledge maps produced, promoted and protected by British Columbia's First Nations, the mineralogical density models constructed by Geoscience BC and its affiliates are laden with designs on the political and economic future of the province. However abstract, these models serve as "economic images" (Schilling 2013) of exploration: formalized, legally regulated representations of mineral potential that are simultaneously pragmatic simplifications of scientific phenomena, as well as beguiling projections of wealth. Mindful of its mandate to act as the avant garde of mineral development in the province, Geoscience BC has sought since its inception to concentrate its work in areas where the timber industry had been the dominant force in local economies. An ongoing infestation of mountain pine beetles, sustained since the late 1990s by a series of unusually hot and dry summers, has damaged or killed 175,000 square kilometers of pine forests throughout the province, an area equivalent to nearly 20 percent of the total land area of the province (Carroll et al. 2003; Kurz et al. 2008). With the long-term health of these forests now in doubt, Geoscience BC's explicit attempts to sell mineral development as a safe route to economic diversification and a more dependable form of income than logging have been met with interest by regional industry advocacy groups like the Northern Development Initiative Trust (Anon. 2007b, 2). Maps displayed in the organization's annual reports and magazines juxtapose regional-scale mineralogical models and prospective scan sites atop the gray and red splotches denoting the dead and dying trees that fill the affected forests, figuratively layering two different possible environmental and economic futures of these regions on the same geographic grid (ibid., 3).

BOUNDARY CHAOS AND "CULTURAL REDISCOVERY"

In the offices of British Columbia's environmental and cultural conservation planning groups like Ecotrust Canada and Landsong Heritage Consulting, future-minded geographic grids enlist a different cast of historical data to promote alternative visions of the future of the province. Both groups emerged during the mid-1990s, a dramatic time for Canadian First Nations, as legal decisions were beginning to challenge the priority of "traditional" land uses over industrial developments and the responsibility of the provincial government to protect these practices. The precedent set by the 1997 Delgamuukw decision and the Nisga'a Final Agreement a year later generated a groundswell of mapping and data collection activities, leading dozens of other First Nations into debate and arbitration, unable to agree on specific financial debts and new territorial boundaries (Penikett 2006, 2–3). In addition to placing the burden of proof for validating these claims on the Nations themselves, the provincial government also set aside money to support Traditional Land Use (TLU) studies among claimants interested in contested land (Olive and Carruthers 1998). Combining archaeological excavations, archival research, and transcriptions and translations of oral histories throughout the province, these well-funded studies quickly produced a veritable "heritage industry," enrolling legal experts, cartographers and academic specialists from a range of historical and ecological disciplines. These studies also stimulated the establishment of more NGOs, adding to an already-crowded landscape of politically minded organizations policing the mining, timber and commercial fishing industries. Even as the methods used to produce these studies grew more consistent, the aims of each First Nation group varied, some planning to use territorial rights to enforce environmental conservation initiatives, with others using them to court mineral developers and to earn increased royalties from timber harvests and sales.

Prior to First Nations corporations' recent efforts to hire their own full-time GIS specialists and archaeologists, consulting groups like Ecotrust and Landsong designed and managed many of the information gathering and assembly steps through which traditional use maps are composed, promoting standardized methods for gathering "cultural data" while establishing a technological system to support and build legitimacy for traditional use mapmaking projects in general. Other consulting groups have worked alongside the treaty proceedings, lobbying provincial lawmakers to recognize and support traditional forest management practices correlated to GIS databanks and online maps (Northern Lights Heritage Service and L. Larcombe Archaeological Consulting 1999). Ecotrust has been particularly active in building infrastructure to extend the use of digital mapping technologies, holding conferences and commissioning pamphlets and studies to promote specific techniques for producing maps and for using them in treaty negotiations (Tobias 2009). In 1998, the group (working jointly with the Gitxsan and the Ahousaht First Nations) launched the Aboriginal

Mapping Network (AMN), an online archive and message board designed to publicize ongoing mapping projects and GIS training and information seminars, share images and articles and connect aspiring cartographers and First Nations activists (Olive and Carruthers 1998). Whereas First Nations advocates are seeking territorial rights in order to better pursue a range of both development and conservation projects, many of the NGOs with whom they work are more explicitly concerned with environmental conservation, or as Ecotrust puts it in much of its promotional literature, with "building the conservation economy" (Schoonmaker, Von Hagen and Wolf 1997). Describing their aims with traditional use maps at a conference in 1998, the directors of Ecotrust's Mapping Office framed the objects of traditional knowledge that they sought to collect and describe as a fund of private capital. "These maps are, in reality, the top layer of complex cultural resource systems. They not only provide essential confirmation of aboriginal title to traditional lands, but also act as educational tools for cultural rediscovery" (Olive and Carruthers 1998).

THE USES OF TRADITION

First Nations have called upon a variety of precedents to support their land claims cases against the provincial and federal governments, but the sheer density of information represented on the maps produced through TLU studies has often been mobilized as a form of proof in its own right. The characteristics of the information generated through these studies, the meaning of "traditional" and the conditions under which such knowledge is produced, however, have been the focus of long-standing legal and academic debates, both for developments in British Columbia and for resource management conflicts elsewhere in northern Canada (Feit 1988; Cruikshank 1998; Laidler 2006). "Traditional ecological knowledge" is a term that gained currency in the early 1990s, when the US National Park Service (USNPS) began to establish guidelines for recognizing and protecting "traditional cultural properties" (Parker and King [1990] 1998; Mihesuah 2000). The term had come into use among Canadian anthropologists and ecologists a few years earlier, first as a practical label for the knowledge produced through prolonged, firsthand observation, typically at a particular location (Berkes, Folke and Gadgil 1995; Berkes 1999; Huntington 2000). According to ecologist Fikret Berkes, one of the early popularizers of the term, "Traditional ecological knowledge is both cumulative and dynamic, building on experience yet adapting to changes. It is an attribute of communities with historical continuity in resource use on a particular land" (1999, 8). Depending upon the rhetorical context, expressions of traditional ecological knowledge may include references to supernatural agencies or explanations of the links between ecological phenomena, social relationships and human health.

In British Columbia and elsewhere in Canada, the term has been used to refer to many forms of knowledge articulated without reference to scientific theory, including observations of flora, relative changes in wildlife abundance and behavior, weather and climate fluctuations, permafrost and soil conditions, and variations in the behavior of sea ice and their related dangers (Nicholas and Andrews 1997; Wohlforth 2004). While Berkes distinguishes these communities as "by and large . . . nonindustrial or less technologically oriented" (1999, 8), in practice, the boundaries between traditional and scientific knowledge are far from clear. In Arctic Canada, the distinction is further complicated by the fact that fieldwork conducted by biologists, climatologists, ecologists and others has long relied on the assistance of Inuit and other indigenous groups for sample collection, navigation and other forms of critical logistical support—in effect, embedding local knowledge within formalized scientific knowledge emerging from situated studies (Cruikshank 1984; Bielawski 2003; Cruikshank 2005). Jean Lave (1996) and Christine Walley (2002) have also taken issue with the reduction of local knowledge to "place" inherent in the designation "traditional," arguing that observational knowledge often circulates widely from sites of production, albeit through different social networks and modes of communication than those employed by professional scientists. Among archaeologists working with universities and with heritage consultancies like Ecotrust and Landsong, the status of oral histories, the ethics of collaboration and the fungibility of knowledge produced through excavations and traditional use studies have already provoked widespread tension and debate (Anyon et al.1997; McGuire 1997; Webmoor 2007).

Equally critical to the substantive content of traditional knowledge is the form in which it is delivered, and, in the eyes of the First Nations elders from whom much of it is collected, it holds its meaning. Uses of traditional knowledge in public policy tend to isolate the empirical details of observations from their more elaborate contexts, gathering this information in "structured interviews quite unlike the usual long . . . narrative[s]" (Bielawski 1995, 223) in which it is usually shared and expressed. Many of these official uses of traditional ecological knowledge also refer to it, or more specifically to the empirical information derived from it, with an acronym: "TEK." Some ecological consultants seeking "wider application of TEK-derived information" in provincial environmental policy acknowledge the problems inherent in the procedural "need to describe TEK in Western scientific terms" (Huntington 2000, 1270) even while categorically distinguishing it from "scientific knowledge" based largely on the manner in which the knowledge is collected.

KNOWLEDGE, POLYGONS AND MAPS

Geographer Nicolas Houde sees the construction and use of TEK-based maps as a double-bind, since the compulsion to defend traditional characterizations of lifestyles leaves many First Nations representatives open to charges of essentializing their own cultures, an accusation echoed by many of the critics of the proprietary battles initiated by USNPS definitions of traditional cultural properties (Mihesuah 2000). Houde comments, "In the public eye, First Nations' legitimacy in negotiating for the co-management of the land partly lies in the existence of located ancestral traditions," since, supposedly, if a "culture does transform over time or moves through space, it is no longer traditional" (2007, 37). While some social scientists and First Nations advocates have fought against the anonymizing conventions of most GIS mapmaking software, personalized and media-rich mapping projects have yet to receive much attention from treaty negotiators or lawmakers.[3] Houde himself first encountered "TEK" in demographic databases and developmental planning maps while working as a policy advisor for a First Nation in Quebec. Thinking "TEK" was merely another obscure governmental acronym, he visually perceived it as an abstract distribution of "discrete entities" and "polygons on maps," realizing only later that the colored shapes marked a wide range of "sites of significance to be protected from logging operations" after taking the time to decode the many additional acronyms used to label the markers (ibid., 36).

Markers generated during the production of traditional use maps have entered environmental policy frameworks through a variety of routes, and TEK-based maps have been used by First Nations to secure rights for both environmental protections and for developmental royalties. Recently, however, the provincial government of British Columbia and a range of cultural conservation groups have sought to expand and standardize digital archives of TEK datasets ever since heritage consultants and First Nations GIS specialists first began conducting TLU studies two decades ago. The provincial government, which provided the original funding for most of these studies, recognized the potential privacy issues associated with creating a centrally controlled database, and originally proposed that TLU grant recipients provide only their "map data, without descriptions of their meaning or significance" for these databases, so that the data could be made "directly accessible for agency planning and for a first stage review of development plans" (Weinstein 1998, np). As First Nations advocates were quick to argue, however, the detailed references associated with anonymized map data could still be made public through applications made through the Freedom of Information and Privacy Act (Olive and Carruthers 1998). Such a possibility raises issues not only of proper translation but also of the political arguments and economic purposes to which unguarded data might be put, and how these themes might "be projected in reports, scholarly journals, or maps" (Houde 2007, 37).

As political scientist Arun Agrawal (1995) has pointed out, the establishment of standardized archives also entails the creation of "an international group of new development professionals, scientifically trained in the latest methods of classification, cataloging, documentation, electronic and physical data storage, and dissemination through publications. Constant attempts to update [these archives] by gathering more information and data . . . will provide purpose and meaning only to a battery of elite data gatherers and analyzers" (Agrawal 1995, 428–429). Whether used to validate claims in treaty negotiations, inform environmental policy or merely exist in salvaged form from "endangered" indigenous cultures (a collecting practice Agrawal derisively refers to as *"ex situ* conservation"), managing and manipulating the maps and datasets produced through traditional use studies is already a technocratic process. While some of the men and women who perform this knowledge-work are themselves members of some of the First Nations directly involved in treaty negotiations, their authority as mapmakers stems as much from their cultural affiliations as it does from the professional associations that promote digital mapmaking practices, and the technical training and computer literacy required to develop GIS expertise.

While some "data" collection projects have fostered fruitful collaborations between technocratic regulators and various First Nations, quite often the maps produced through these studies are sent to other regulatory contexts in which they are expected to speak for themselves. If a particular set of markings earns environmental protection, for instance, their presence alone can render certain kinds of investigation and technical description off limits. Like the visual artifacts of geostatistical modeling, these markers are encountered by federal and provincial regulators and planners not as raw geological strata or layers of locally cultivated traditions, but as visual languages and abstract iconographies. Also like the models and datasets offered up by Geoscience BC, TEK icons are often taken up by different groups and reassembled with other kinds of markers into still more maps; once transposed, the geographical coordinates of the markers are sometimes all that remain of the original content and meaning of the icons.

Abstraction and formalization shaped the conduct of cartographers and the reception of maps long before the advent of digital mapmaking, but few paper maps invite the same degree of improvisational reconstruction as the mineralogical models and traditional use maps currently competing for the attention of potential investors and treaty negotiators in British Columbia. Much as the original sketches and charts of the British Columbia Geological Survey were shaped by the routes of access and developmental ideologies of the province's first geologists and mining engineers, GIS, digital cartography, geostatistical modeling and the other tools that make up geoinformatics depend upon the institutional infrastructure and technocratic work that keeps online databases publicly accessible and tacitly federated. As Geoscience BC and the Aboriginal Mapping Network use their maps to draw more investors and First Nations into land claims negotiations and developmental debates, their standards, tools and necessary omissions are leaving a considerable imprint upon the political and economic visions that these groups are projecting for the province. In the age of computerization, British Columbia's proliferating and ambiguous futures reside in the visual language of maps.

NOTES

1. Unless otherwise noted, the material in this article is based on information obtained during interviews conducted with geologists, exploration company officials and geostatistics consultants in Alaska, Wyoming and British Columbia between June and August 2011.
2. A note on terminology: the Canadian social science and regulatory literature tends to use "aboriginal" most frequently as a general label for multiple groups, while the terms "indigenous," "Indian" and "Native American" are more common in the US. As a legal status marker, the term "First Nations" applies to non-Inuit, non-Métis (mixed First Nations/white parentage) in the Canadian provinces.
3. One such system, dubbed the "Geographical Valuation System," or GVS, was designed jointly by the University of Northern British Columbia and the Halfway River First Nation; literature on the system asserts that the GVS was designed to be useful and accessible to all interested First Nations (Elliot 2008).

REFERENCES

Agrawal, Arun. 1995. "Dismantling the Divide between Indigenous and Scientific Knowledge." *Development and Change* 26 (3): 413–439.

Anon. 2007a. "Mountain Pine Beetle Initiatives: Unlocking BC's Potential with Geoscience!" *Geoscience BC's Explorer: Annual Report 2007*: 2.

Anon. 2007b. "The Northern Trust: Geoscience BC's Partner in Economic Diversification of the MPB Area". *Geoscience BC's Explorer: Annual Report 2007*: 3.

Anon. 2011a. "Geoscience BC" (personal communication, June 20).

Anon. 2011b. "Mineral Development Office of British Columbia Bureau of Energy Mines, and Petroleum Resources" (personal communication, June 20).

Anon. 2011c. "From Sea to Sea: A Brief Introduction to Canada." *Engineering and Mining Journal: Report on Mining Industry in British Columbia and Yukon, July–August*: 53.

Anyon, Roger, T. J. Ferguson, Loretta Jackson, Lillie Lane and Philip Vincenti. 1997. "Native American Oral Tradition and Archaeology: Issues of Structure, Relevance, and Respect." In *Native Americans and Archaeologists: Stepping Stones to Common Ground*, edited by Nina Swidler, Kurt Dongoske, Roger Anyon and Alan Downer, 77–87. Walnut Creek, CA: AltaMira.

Bankes, Nigel. 1998. "Delgamuukw, Division of Powers and Provincial Land and Resource Laws: Some Implications for Provincial Resource Rights." *U. Brit. Colum. Law Review* 32: 317–352.

Barman, Jean. (1991) 2007. *The West beyond the West: A History of British Columbia*. 3rd ed. Toronto: University of Toronto Press.

Basso, Keith H. 1996. *Wisdom Sits in Places: Landscape and Language among the Western Apache*. Albuquerque: University of New Mexico Press.

Berkes, Fikret. 1999. *Sacred Ecology: Traditional Ecological Knowledge and Resource Management*. Washington, DC: Hemisphere.

Berkes, Fikret, Carl Folke and Madhav Gadgil. 1995. *Traditional Ecological Knowledge, Biodiversity, Resilience and Sustainability*. Stockholm: Beijer International Institute of Ecological Economics.

Bielawski, Ellen. 1995. "Inuit Indigenous Knowledge and Science in the Arctic." In *Human Ecology and Climate Change: People and Resources in the Far North*, edited by David L. Peterson and Darryll R. Johnson, 219–227. London: Taylor and Francis.

Bielawski, Ellen. 2003. "'Nature Doesn't Come as Clean as We Can Think It': Dene, Inuit, Scientists, Nature and Environment in the Canadian North." In *Nature across Cultures: Views on Nature and the Environment in Non-western Cultures*, edited by Helaine Selin, 311–327. London: Kluwer Academic.

Bourdieu, Pierre. 1990. "The Kabyle House or the World Reversed." In *The Logic of Practice*, translated by R. Nice, 271–283. Stanford, CA: Stanford University Press.

Braun, Bruce. 2000. "Producing Vertical Territory: Geology and Governmentality in late Victorian Canada." *Cultural Geographies* 7 (1): 7–46.

Burnett, D. Graham. 2001. *Masters of All They Surveyed: Exploration, Geography, and a British El Dorado*. Chicago: University of Chicago Press.

Carroll, Allan L., et al. 2003. "Effect of Climate Change on Range Expansion by the Mountain Pine Beetle in British Columbia." In *Mountain Pine Beetle Symposium: Challenges and Solutions* (Information Report BC-X-399), edited by Terence L. Shore, 223–232. Kelowna, BC: Natural Resources Canada.

Chambers, Kimberley J., Jonathan Corbett, C. Peter Keller and Colin J. B. Wood. 2004. "Indigenous Knowledge, Mapping, and GIS: A Diffusion of Innovation Perspective." *Cartographica* 39 (3): 19–31.

Chapin, Mac, Zachary Lamb and B. Threlkeld. 2005. "Mapping Indigenous Lands." *Annual Review of Anthropology* 34: 619–638.

Claims Registry. 2012. *British Columbia Ministry of Energy and Mines*. Accessed January 8, 2013. https://www.mtonline.gov.bc.ca/mtov/home.do. Last accessed 4 November 2013.

Corburn, Jason. 2005. *Street Science: Community Knowledge and Environmental Health Justice*. Cambridge, MA: MIT Press.

Cruikshank, Julie. 1984. "Oral Tradition and Scientific Research: Approaches to Knowledge in the North." In *Social Science in the North: Communicating Northern Values*. Occasional Publication No. 9. Ottawa: Association of Canadian Universities for Northern Studies, 3–24.

Cruikshank, Julie. 1998. *The Social Life of Stories: Narrative and Knowledge in the Yukon Territory*. Lincoln: University of Nebraska Press.

Cruikshank, Julie. 2005. *Do Glaciers Listen?: Local Knowledge, Colonial Encounters, and Social Imagination*. Seattle: University of Washington Press.

Dosemagen, Shannon, Jeff Warren and Sara Wylie. 2011. "Grassroots Mapping: Creating a Participatory Map-Making Process Centered on Discourse." Issue 8: Grassroots Modernism. *Journal of Aesthetics and Protest*.

Edney, Matthew H. 1997. *Mapping an Empire: The Geographical Construction of British India, 1765–1843*. Chicago: University of Chicago Press.

Elliot, Nancy J. 2008. *Including Aboriginal Values in Resource Management through Enhanced Geospatial Communication*. London, ON: Bibliothèque et Archives Canada.

Feit, Harvey. 1988. "Self-Management and State-Management: Forms of Knowing and Managing Northern Wildlife." In *Traditional Knowledge and Renewable Resource Management in Northern Regions*, edited by Milton M. R. Freeman and Ludwig N. Carbyn, 72–91. Edmonton: Boreal Institute for Northern Studies.

Fondahl, Gail. 2006. "Increasing Indigenous Participation in Resource Management in the North: Research Opportunities and Barriers." Paper presented at the Fourth Open Assembly of the Northern Research Forum (NRF) "The Borderless North." Tornio, Finland. October 5–8.

Gell, Alfred. 1985. "How to Read a Map." *Man* 20 (2): 271–286.

Grek-Martin, Jason. 2011. "Canada and the Colonization of the Canadian West in the Late 19th Century." PhD diss., Queen's University.

Harley, J. Brian. 1992. "Deconstructing the Map". In *Writing Worlds: Discourse, Texts and Metaphor in the Representation of Landscape*, edited by Trevor J. Barnes and James S. Duncan, 231–247. London: Routledge.

Helmreich, Stefan. 2011. "From Spaceship Earth to Google Ocean: Planetary Icons, Indexes, and Infrastructures." *Social Research* 78 (4): 1211–1242.

Houde, Nicolas. 2007. "The Six Faces of Traditional Ecological Knowledge: Challenges and Opportunities for Canadian Co-management Arrangements." *Ecology and Society* 12 (2): 34–51.

Howard-Grenville, Jennifer, Jennifer Nash and Cary Coglianese. 2008. "Constructing the License to Operate: Internal Factors and Their Influence on Corporate Environmental Decisions." *Law and Policy* 30 (1): 73–107.

Huntington, Henry P. 2000. Using Traditional Ecological Knowledge in Science: Methods and Applications. *Ecological Applications* 10 (5): 1270–1274.

Kurz, Werner, Caren Dymond, Graham Stinson, Greg Rampley, Eric Neilson, Allan Carroll, Tim Ebata and Les Safranyik. 2008. "Mountain Pine Beetle and Forest Carbon Feedback to Climate Change." *Nature* 452 (7190): 987–990.

Laidler, Gita J. 2006. "Inuit and Scientific Perspectives on the Relationship between Sea Ice and Climate Change: The Ideal Complement?" *Climatic Change* 78 (2–4): 407–444.

Lave, Jean. 1996. "The Savagery of the Domestic Mind." In *Naked Science: Anthropological Inquiry into Boundaries, Power, and Knowledge*, edited by Laura Nader, 87–100. London: Routledge.

McGuire, Randall. 1997. "Why Have Archaeologists Thought Real Indians Were Dead and What Can We Do About It?" In *Indians and Anthropologists: Vine Deloria Jr., and the Critique of Anthropology*, edited by Thomas Biolsi and Larry Zimmerman, 63–91. Tucson: University of Arizona Press.

Mihesuah, Devon A., ed. 2000. *Repatriation Reader: Who Owns American Indian Remains?* Lincoln: University of Nebraska Press.

Mira Geoscience. 2011. *Regional 3D Inversion Modelling of Airborne Gravity and Magnetic Data: QUEST-South, BC, Canada* (Report #2011–14). Vancouver, BC: Geoscience BC.

Nicholas, George P., and Thomas D. Andrews, eds. 1997. *At a Crossroads: Archaeology and First Peoples in Canada*. Burnaby, BC: Archaeology Press.

Northern Lights Heritage Service and L. Larcombe Archaeological Consulting. 1999. *Criteria and Indicators for Naturalized Knowledge: Framework and Workshop Proceedings—Prepared for the Prince Albert Model Forest and the Naturalized Knowledge Working Group*. Winnipeg: Northern Lights Heritage Service.

Olive, Caron, and David Carruthers. 1998. "Putting TEK into Action: Mapping the Transition." Paper presented at Ecotrust Canada meeting "Bridging Traditional Ecological Knowledge and Ecosystem Science." Flagstaff, AZ. August 13–15, 1998.

Oreskes, Naomi. 2007. "From Scaling to Simulation: Changing Meanings and Ambitions of Models in the Earth Sciences." In *Science without Laws: Model Systems, Cases, and Exemplary Narratives*, edited by Angela N. H. Creager, Elizabeth Lunbeck and M. Norton Wise, 93–124. Durham, NC: Duke University Press.

Oreskes, Naomi, Kristin Shrader-Frechette and Kenneth Belitz. 1994. "Verification, Validation, and Confirmation of Numerical Models in the Earth Sciences." *Science* 263 (5147): 641–646.

Orlove, Ben 1991. "Mapping Reeds and Reading Maps: The Politics of Representation in Lake Titicaca." *American Ethnologist* 18 (1): 3–38.

Parker, Patricia L., and Thomas F. King. (1990) 1998. *Guidelines for Evaluating and Documenting Traditional Cultural Properties*. Washington, DC: US Department of the Interior and National Park Service.

Peluso, Nancy Lee. 1995. "Whose Woods Are These?: Counter-Mapping Forest Territories in Kalimantan, Indonesia." *Antipode* 27 (4): 383–406.

Penikett, Tony. 2006. *Reconciliation: First Nations Treaty Making in British Columbia*. Vancouver: Douglas & McIntyre.

Rankin, William. 2011. "After the Map: Cartography, Navigation, and the Transformation of Territory in the Twentieth Century." PhD diss., Harvard University.

Schilling, Tom. 2013. "Uranium, Geoinformatics, and the Economic Image of Uranium Exploration." *Endeavour* 37 (3): 140–149.

Schoonmaker, Peter K., Bettine Von Hagen and Edward C. Wolf. 1997. *The Rain Forests of Home: Profile of a North American Bioregion*. Washington, DC: Island Press.

Tobias, Terry N. 2009. *Living Proof: The Essential Data-Collection Guide for Indigenous Use-And-Occupancy Map Surveys*. Vancouver, BC: Ecotrust Canada.

VanDine, Douglas F., Hugh W. Nasmith and Charles F. Ripley. 1993. "The Emergence of Engineering Geology in British Columbia: 'An Engineering Geologist Knows a Dam Site Better!'" Vancouver: British Columbia Ministry of Energy and Mines.

Wainwright, Joel, and Joe Bryan. 2009. "Cartography, Territory, Property: Postcolonial Reflections on Indigenous Counter-Mapping in Nicaragua and Belize." *Cultural Geographies* 16 (2): 53–178.

Walley, Christine J. 2002. "'They Scorn Us Because We Are Uneducated': Knowledge and Power in a Tanzanian Marine Park." *Ethnography* 3 (3): 265–298.

Webmoor, Timothy. 2005. "Mediational Techniques and Conceptual Frameworks in Archaeology: A Model in 'Mapwork' at Teotihuacan, Mexico." *Journal of Social Archaeology* 5 (1): 52–84.

Webmoor, Timothy. 2007. "The Dilemma of Contact: Archaeology's Ethics-Epistemology Crisis and the Recovery of the Pragmatic Sensibility." *Stanford Journal of Archaeology* 5: 224–246.

Weinstein, Martin S. 1998. "Sharing Information or Captured Heritage: Access to Community Geographic Knowledge and the State's Responsibility to Protect Aboriginal Rights in B.C." Paper presented at Ecotrust Canada meeting "Crossing Boundaries: 7th Conference of the International Association for the Study of Common Property." Vancouver, BC. June 9–14, 1998.

Winichakul, Thongchai. 1994. *Siam Mapped: A History of the Geo-body of a Nation*. Honolulu: University of Hawaii Press.

Winsberg, Eric. 1999. "Sanctioning Models: The Epistemology of Simulation." *Science in Context* 12 (2): 275–292.

Winsberg, Eric. 2010. *Science in the Age of Computer Simulation*. Chicago: University of Chicago Press.

Wohlforth, Charles. 2004. *The Whale and the Supercomputer: On the Northern Front of Climate Change*. New York: North Point Press.

Zaslow, Morris. 1975. *Reading the Rocks: The Story of the Geological Survey of Canada, 1842–1972*. Toronto: Macmillan.

4 Redistributing Representational Work

Tracing a Material Multidisciplinary Link

David Ribes

INTRODUCTION

Traditionally, we understand multidisciplinary links to be the work of human relationships across domains of heterogeneous expertise—for example, bringing together biologists and physicists in order to understand genetics (Kay 2000), or a collaboration between computer scientists and geologists as a means of framing a new understanding of geoinformatics (Ribes and Bowker 2009). This human focus directs our attention to questions of communication, shared language or diverging understandings (Galison 1999; Jeffrey 2003). From this perspective, technology plays a supportive role in human collaborations—facilitating communication across domains, time and/or space (Olson, Zimmerman et al. 2008). Yet multidisciplinarity does not rely solely on a human-to-human links: Frequently, technology plays a leading role in collaboration across disciplines. In this chapter, I focus on how technology becomes the multidisciplinary link, sustaining relations across domains of scientific expertise that are not centered on human-to-human ties (Ribes, Jackson et al. 2012).

While produced through multidisciplinary collaboration, certain representational technologies can become relatively autonomous of the initial human-to-human relationships responsible for their creation. Through visual output, multidisciplinarity can also be structured through representation tools—technologies that continue to carry a history of their development over time. By tracing the production of visualization systems, such as those used in the sciences and medicine, we as analysts must also travel across multiple disciplinary boundaries. This chapter is one such journey: the exploration of a distinct multidisciplinary link, in the end tied together not by the human-to-human collaboration of heterogeneous experts but by the development, and thereafter use, of technology.

Multidisciplinary relationships sustained through technology restructure how we should think about knowledge production. In visualization studies we have come to consider representational technique as tied to scientific knowing. For example, in his classic study of visual representation in geology, Martin Rudwick claimed that we should take the "development of the visual language of geology not only for the way that it gradually enabled the concepts of a new science to be more adequately expressed, but also as a reflection of the growth of a self-conscious community of geological scientists" (Rudwick 1976, 151). Visualizations, from this perspective, are a shorthand for epistemic commitments. But with the redistribution of representational work—by "outsourcing" of the methods of representation to a group external to the domain—we are witnessing the reconfiguration of scientific relations of knowledge production (Latour 1986; Hutchins 1995; Suchman 2007). Programmers and their software output now mediate between the practices of data generation and their analysis as visual images.

This chapter is structured as follows. I first introduce today's visualization researchers and the double trajectory of their research output: These scholars are both publishing research findings and producing visualization software. Using a case of visualization research that seeks to "capture" existing artistic techniques to inform the design of novel visualization tools, I trace the work of "Marie" as it follows two distinct trajectories: 1) published findings as academic research and 2) the production of data visualization software used in the sciences. In the first trajectory—publication—Marie reflects critically upon her research design, as is common with experimental research. That is, in her papers Marie identifies flaws or weaknesses in her empirical research and experimental approach. However, in the second trajectory her research is built into visualization tools that circulate across disciplinary boundaries. Scientists using these visualization tools in their research have little, if any, knowledge of the experimental problems that Marie encountered in the tool's development. It is the gap between the first and second trajectories that interests me: The methodological problems outlined within her published research are difficult to identify in her visualization software. In the second half of this chapter I follow Marie's code itself as she translated it across multiple sites of application. While the algorithms described in this chapter were initially intended for visualizing cancer treatment regimes, their "social life" did not end with a single application. Marie works with many disciplines, and her visualization tools travel with her. Findings from visualization research and code from visualization software were reassembled for use as data visualization software in new settings. With each such translation across disciplines, the origins of the production of software tools become increasingly difficult to discern.

While we usually consider scientists to be the arbiters of their own representational techniques, in Marie's case I find a distinct rupture between tool development and its use. Visualization tools not only sustain the multidisciplinary relationship between visualization scholars and scientists but also hide, or render invisible, the problems faced in their development from those who use these tools.

VISUALIZATION RESEARCHERS

In 1963, Ivan Sutherland produced what is often cited as the first computer graphical user interface (GUI) as part of his dissertation, entitled "Sketchpad: A Man-Machine Graphical Communications System." Sketchpad, a system for computerizing the practice of drawing and design, was displayed on a monitor utilizing a physical interface of a light pen, switches and knobs. Sutherland innovated techniques such as memory structures in order to store objects; the rubber banding of lines; the ability to zoom in and out on the computer display; and even the ability to make automatically rendered lines, corners and joints. These techniques—now familiar to any user of low-end drawing or painting programs—also remain the staples of all forms of computer-aided design (CAD).

Today, fifty years after Sutherland's innovation, his "new science" is in full swing. An entire subdiscipline of computer science (CS) and information technology (IT) has emerged dedicated to the production of digital visualization tools. These programmers, or at least their software products, now mediate between the activities of generating data and producing knowledge; this software operates in the interstices between data collection and the dissemination of scientific knowledge. In other words, visualization researchers create the tools for translating specific information into meaningful images for those in the sciences and engineering. Visualization researchers—or more specifically, the tools they build—stand between "raw data" (Ribes and Jackson 2013) and their meaningful interpretation.

Whether the design practices explored in this chapter are considered a science of computing or a sort of engineering, creators of visualization tools are not programmers alone, but also empirical researchers. Visualization experts are trained in many fields (e.g., engineering, mathematics, computer science, information science, etc.) and draw from a wide variety of heterogeneous disciplinary traditions, as evidenced by their idiosyncratic backgrounds. In the US, thousands of such researchers congregate annually at conferences on interface design (e.g., CHI) or computer graphics (e.g., SIGGRAPH), demonstrating their findings, doing proof-of-concept demonstrations or demonstrating commercial visualization tools. They may participate in fields such as human-computer interaction (HCI) or publish in specialized visualization journals. Their success in these fields is often tied to publication *and* to producing novel tools for visualization.

Visualization researchers may very well have a professional interest in the psychology of perception, the biology of the eye and brain or even techniques of illustration. For them, each of these activities or interests can inform how humans see and render imagery. In this sense, visualization researchers should be understood as sitting at the intersection of two diverging trajectories: 1) empirical research contributing to theoretical understandings of perception,

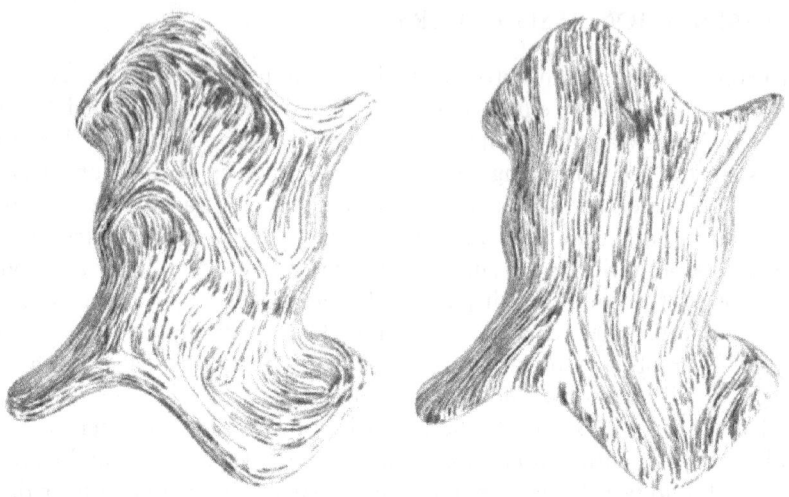

Figure 4.1   Two examples of Marie's work in the application of texture-mapping to a single surface. Which is more *effective*? © 2003 IEEE. Reprinted with permission from *IEEE Transactions on Visualization and Computer Graphics*.

visualization, image interpretation and human-computer interaction and 2) efforts to produce novel visualization software.

The software products of this second trajectory may come to be a tool for researchers in another scientific domain as they conduct data analysis. It is this transition that interests me. In this chapter I trace those visual rendering techniques that were once the *object of visualization research* and thereafter became the *tools for data representation in sciences and medicine*.

This chapter highlights one computer scientist, "Marie," who conducts research for the purpose of creating visualization tools. I do not claim that Marie "stands-in" as a representative for her field as a whole; instead, I chose her specifically for the range of methodological approaches she applies to visualization research, and to illustrate the unique connections she builds between techniques of visual rendering and contemporary scientific practice.

Marie's inspiration is the sophisticated use of technique in the arts. Her research focuses on applying insights from visual perception, art and illustration to the design of more effective techniques for data visualization (See Figure 4.1). She is in contact with disciplines as diverse as psychology, biology, art history, statistics and, of course, her own domain of computer science; her methodologies reflect this particular blend of multidisciplinarity. As Marie explains, her interests are grounded in *formalizing the knowledge and practice of arts and design and then incorporating this knowledge within information systems*. These inspirations come to be "captured" within the systems she designs.

The computer sciences do not often teach empirical methodologies, and in the case of Marie, it is largely her concordant training in experimental psychology that informs her empirical research. One of her main foci has been the study of texture's effect on the perception of shape and the use of texture-synthesis for shape representation, or, more specifically: the study of texture perception and classification for texture-synthesis in multivariate data visualization. In other words, how do you represent data as surfaces in order to maximize perception of topology?

Marie is both a scientist and an applied researcher: Through shared funding or contract work, she collaborates with biological and medical researchers to develop tools that support their work. In the first half of this chapter, I follow the development of one tool initially intended to represent the time lines of radiation exposure in cancer treatment. However, as we will see in the second half of this paper ("Mobile Code"), Marie also uses these same findings and software creations to produce tools for very different purposes: specifically, software that can represent the structure of veins and nanoparticle flows. Her tools travel with her as she works across disciplinary boundaries and domain applications, as well as across federal research grants and private collaborations.

Rather than positing "influence," "cultural diffusion" or loose borrowing, I follow the chain of productive work from analog to digital media, through research practice, and then programming. It is this productive work that links such diverse domains as design, HCI, visualization, laboratory practice and scientific publication. In the software products of visualization work, multidisciplinary links are maintained through the work of automated material relations.

CAPTURING THE ARTS AND DESIGN

This section outlines how artistic and illustrative techniques come to be scientific research objects, showcasing how Marie subjects them to experimental conditions. Marie reveres artists' ability to effectively render surface. She takes artists, as well as art and art history, as an enormous repository of informal knowledge to be systematized: "Observation of the practices of artists and illustrators provides a rich source of inspiration for the design of more complex and possibly more intuitively appealing methods for translating data into pictures" (publication).

As a computer scientist, Marie tasks herself with placing artists' informal, tacit or embodied knowledge within information systems. Drawn from an oral conference presentation, in the following excerpt she neatly encapsulates the goal of automating visualization:

> I went to St. Paul's Library and I was looking at these textbooks on fabric design and quilting and looking at these incredible pictures where

they'd taken fabrics of different colours and textures and they had woven them into these beautiful artworks. And they were all different *and they all worked.* And I was trying to think, why is it that some things work and some things don't work? And I was trying to think can we measure this mathematically, so that people like me, who have a little bit of intuition but maybe not a lot of intuition, can maybe get a hand in trying to figure out something that works, how to choose something that works? But the problem is right now I'm having trouble figuring out what kind of statistics to use, and what the correlation is. (Conference presentation, emphasis in original)

To sum up, she took inspiration from the adept use of visual techniques used in art—something she viewed as an artist's sheer effectiveness in conveying a message. Subsequently, she sought to harness these skills to serve not only her own nonartistic practices but also those of her collaborators in the sciences.

However, capturing the arts is a complex endeavor. In this process art, art history and artists themselves become objects of the scientific gaze. Art and design techniques were the scientific objects on which Marie intervened. Art was made an analyzable object by subjecting it to the various techniques available to empirical research. Marie worked to make artistry (textures) researchable by embedding them within a distinct *experimental system*—an assembly of the research object, scientific discourse, instrumentation and practice (Rheinberger 2000). Experimental systems allow scientists to intervene with, to shape and to represent scientific objects. They embed scientific objects into a broader field of material scientific culture and practice, of instrumentation and inscription devices.

Marie used the methods of perceptual psychology for her studies. She brought art into an experimental context: "Research in perceptual psychology provides a rich source for insight into the fundamental principles underlying the creation of images that can be effectively interpreted by the human visual system" (publication). Inspired by texture use in medieval art, Marie selected a variety of sample patterns—these became "stimuli" in the language of perceptual psychology—and mapped them onto computer-generated surfaces. She took physical samples of medieval tapestries and digitized them using a high-fidelity scanner, and then created simple algorithms that allowed the strategic placement of these patterns on topological surfaces:

The stimuli that we used in our experiments were cropped images of the front-facing portions of textured level surfaces rendered in perspective projection using a hybrid renderer . . . that uses raycasting . . . together with a Marching Cubes algorithm . . . for surface localization. (Publication)

The particular topological surfaces she chose to represent were "three dimensional dose distributions calculated for a radiation therapy treatment plan" (publication)—these are a medical representation for determining a time line of radiation exposure levels as part of cancer treatment. The particular set of experiments she described here was conducted in conjunction with funding to develop this application. The surfaces were selected because rendering software would "typically be" this sort of visualization:

> We chose to use the radiation data as our test bed, rather than a more restricted type of analytically-defined surface, because this data is typical of the kind of data whose shape features we seek to be able to more effectively portray through the use of surface texture. (Publication)

Marie generally works on data that, when visually represented, form a plane rather than a discrete object, or, as she puts it, where "shape-edge is not an available cue" (publication). Any object seen up close does not have edges; to perceive its shape, other cues must serve to render form, such as texture. See, for example, Figure 4.2, which includes a shaped surface with no edges. Thus, while Marie's application is specific (radiation exposure time lines), she was interested in a more general outcome (representing data without shape-edge cues). This focus was Marie's basis for generalizing her rendering tools across specific applications, a topic I return to later in the paper ("Mobile Code").

To test the effectiveness of texture in facilitating the perception of a shape without edge-cues, Marie and her research team first recruited a population of thirty "properly controlled" subjects. In two-hour-long trials for each of the thirty participants, subjects were asked to observe six computer-generated topological surfaces, each coated with digital "probes" that could be adjusted by "pulling on its handle until the circular base appear[s] to lie in the tangent plane to the surface at its central point and the perpendicular extension appear[s] to point in the surface normal direction" (publication; see Figure 4.2). Each surface was texture-mapped using various rendering algorithms drawn from real-life samples, such as wall hangings, paintings, sketches or even woven cloths.

The experiment yielded quantitative statistical results measuring the accuracy of the subject's ability to perceive topology. To understand and, most importantly, pinpoint the most effective textures, Marie compared the data, finding trends and generalizations across both the subjects and the various texture-maps (see Figure 4.2).

The results of Marie's experiment supported her "hypothesis that texture pattern anisotropy impedes surface shape perception in the case that the direction of the anisotropy does not locally follow the direction of greatest normal curvature" (publication). Or, put somewhat more straightforwardly, Marie found that human subjects could distinguish shape best when lined patterns are used, and specifically, if those lines went against the grain of curvature.

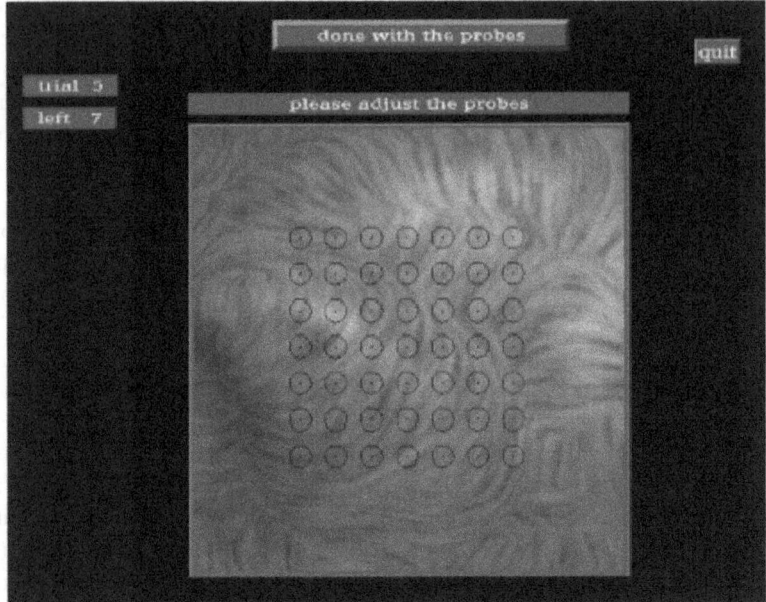

Figure 4.2   A texture-map used in one of Marie's experimental systems. Forty-nine "probes" appear on the map, as depicted by the dots within the circles. The user can adjust these to line up his or her perception of the topology. The surfaces themselves acted as 3-D renderings of radiation therapy exposure treatments. Reprinted with permission from *SPIE Proceedings.*

One may ask, "How could art be captured? Its qualities are sublime." But from the designer's perspective, this is a somewhat Romantic formulation. In fact, Marie acknowledges that it is not the ineffable qualities of "art" that are necessarily captured in automated systems, but rather the representational techniques that make them up:

> Visualization differs from art in that its ultimate goal is not to please the eye or to stir the senses but, far more mundanely, to communicate information—to portray a set of data in a pictorial form that facilitates its understanding. (Publication)

Scientific visualization does not seek *only* to "represent reality"; this is but one of its many concerns (c.f. Lynch and Edgerton 1988). "You need to be able to see something new, or even something easier, and sometimes that even means seeing something *wrong*—because domain experts don't just see my visual they also see everything else they know." Marie's visualizations seek to allow certain *kinds* of seeing that inform, while at the same time sustain, an orthogonal concern for reality. It is the interpretation of data (i.e., human perception of topology) that she seeks to facilitate: "The ultimate success of a visualization can be objectively measured in terms of the extent to which it proves useful in practice. But to take the narrow view that aesthetics don't matter is to overlook the complexity of visual understanding" (publication). Rather than capturing artistic technique, what interests these researchers is building a close tie between human perceptual systems and data (Tufte 1986).

Two Streams of Output: Publishing and Application

Marie, in addition to being interested in capturing art and design practices, is also application oriented. The study of the arts—textures and their relation to human perception—eventually comes to inform the construction of visualization tools for science and engineering research. Marie's research feeds into academic publication *and* the production of software for science, medicine and engineering, thus producing two distinct streams of output. Beyond those interventions to make artistry accessible to the scientific gaze described earlier, findings must be embedded within the software suites that are intended to render data as image—what the computational worlds call *applications*. It requires a very specific articulation, or what programmers call *coding*, to render the findings of research into machine language. Thereafter Marie's research was "applied," and her output is executable as a software algorithm.

Marie was capturing artistic technique *and* closely tuning a product to an existing stream of scientific practice. The study discussed earlier informed the design of a tool to visualize radiation exposure time lines for cancer patients. Marie worked closely with her various collaborators to ensure the particular visualization tool would be useful—that is, that it functioned with their current operating platforms, or that the interface achieved a balance between configurable and accessible.

Transitions between capture and coding have often been complicated for Marie. A scientific object, here art and design practice, may resist particular techniques of capture and representation. Marie encountered difficulties in producing "clean results" from her perceptual experiments. However, such problems were still several steps removed from application; they were still concerned with human perception:

> The root of our difficulties was that too many of the points on our surfaces were too near to being parallel with the image plane. In numerous incidences the angular deviation in tilt was degenerate, because the estimated normal projected to a single point, and it was not clear how to appropriately handle these cases. We could not simply exclude these samples from our error calculations, because their occurrence was not uniform but tended to predominate in "bad texture" conditions, where the cues to shape were inadequate and subjects reverted to the default assumption that the surface lay in the plane of the image, or subjects

simply gave up in frustration and left the probes untouched at their default original positions. (Publication)

To rephrase: The grid of probes that Marie set upon the texture-map created a flat plane relative to the computer monitor (see Figure 4.2). In the experimental trials, portions of the texture-map appeared to run parallel to this plane, confusing the subjects who then often left the probes unadjusted. Because the problem was exacerbated in those cases that use poor topology emphasizing textures (the *object* of Marie's study), she was unable to simply remove these outliers. Marie's solution was to abandon traditional quantitative technique in favor of less accepted statistical methods, which nevertheless yielded poor results: "We therefore reluctantly decided to break with tradition and simply use as an error metric the angle in $\mathfrak{R}^3$ between the estimated normal direction specified by the probe and the true surface normal direction at the probe center" (publication). Marie described this statistical technique as a "fallback position" that is notably weaker: "If you are a statistician, or experimentalist, you can read between the lines of what it means to use this [error metric]." Only by careful massaging of the data, through analytic approaches, and by tailoring of subjects' responses did she achieve relatively clean results for this experiment.

Here we can begin to trace the two distinct trajectories for Marie's findings: publication and application. It is a well-established tradition within experimental science, when writing up a paper for publication, to present both the weaknesses in design and avenues for future research. In this tradition, Marie provided a critical analysis of her own work in her papers. Her critiques were of the effects of measurement/probing on the subjects, and the limited use of textures:

> Unfortunately, we neglected to recognize, before beginning the experiments, that our decision to place the probes at exactly evenly spaced intervals over a rectangular grid would interfere with observers' ability [to] perceive all of the probes as lying in the surface at the same time, due to violation of the generic viewpoint assumption. (Publication)

Marie divulged to her readers that the probes placed in a regular grid pattern on the "surface" of the texture-mappings—intended to measure subjects' perception of topology—are almost certainly contributing to the perception of shape. Secondly, the patterns used were themselves relatively simple: six variations that map lines in relation to topology. Finally, Marie admitted that her statistical analysis may not have been up to the task of determining clear results: The statistical significance was positive but relatively low, and she had not yet devised a sophisticated analysis of individual mean results. From a scientific viewpoint Marie noted that there is a great deal of room to devise a more sophisticated experimental design for the development of more complex mapping algorithms and that current statistical techniques were inadequate for the task.

And yet her experimental system produced results. As we have seen, despite complications, she still felt able to conclude that "texture pattern anisotropy impedes surface shape perception in the case that the direction of the anisotropy does not locally follow the direction of greatest normal curvature" (publication). Marie *did* develop a series of texture-mapping algorithms. She also invested the time and effort to experimentally test which of these algorithms produced a maximization of topology identification accuracy in subjects. But her data were admittedly flawed, and there were clear avenues for future research. Did Marie wait for more definitive results before embedding her work in computer programs? In this case, no.

In a practice-driven field like human-computer interface, where career rewards are tied to production as much as publication, Marie chose to develop applications from her visualization research data. Moreover, her experimental work was tied to a particular usage: She was building visualization devices to assist medical practitioners in planning cancer therapy schedules—this was a professional relationship between medicine and computing that operated on a schedule independent of refereed publication. In scientific research it is not rare for the results of an experiment to remain inconclusive, but in her software engineering practice Marie must occasionally forgo maximizing understanding of visual acuity in order to satisfy production requirements—that is, make a useable tool.

There is a split between Marie's academic publishing and system design. She is contributing to a body of scholarly work by publishing in the refereed journals of a community of experimental perceptual psychologists. This work is reviewed and critiqued, revised and resubmitted. Her methods are open to evaluation, and Marie made public a self-critique of her experimental approach. But Marie is *also* contributing to a repertoire of technology, to a software visualization program. Marie incorporated her findings into an already existing visualization suite; this program will now automatically determine what texture pattern to overlay on a topology to maximize visual acuity.

The trajectory of this scientific object– texture perception—concluded with its introduction to a new experimental system. Her visualization tool is now used by medical scientists to analyze data and make determinations about treatment regimens. The end result of the automation of art visualization is the background incorporation of these techniques within computer applications. Artistic technique and design-practice were delegated to computer applications and incorporated within scientific and engineering practice in order to render data. Marie's textures from a renaissance wall-hanging, once her object of study, were translated into scientific findings intended to become the tools of scientific data visualization. Historian of science Hans-Jörg Rheinberger points out that "*things*" cease to be scientific objects as they lose their recalcitrance, their resistance to manipulation and predictability. Once they are *sufficiently* malleable to scientific practice they take on a new life, which he describes as "both marginal and central" (Rheinberger 2000, p.275). In the case Rheinberger explores of cytoplasmic particles, these entities became marginal in that they were less the object of scientific scrutiny and investigation and instead became newly central as the platform for further scientific research. That is, cytoplasmic particles became the tools for investigating new phenomena.

This is very much the case in the automation of visualization, where a particular representational schema is initially the object of research for a computer scientist but later, as it becomes incorporated into visualization tools, becomes the platform for research within another science. In this transition, the various uncertainties about this mode of texture representation are largely left behind. The experimental system remains only in traces, all but invisible to a user of the visualization software, and this original output itself becomes a tool for scientific or design visualization, central as another experimental system in the research projects of a domain science.

Marie's research work was simultaneously published in refereed journals, but was also placed in circulation through their designs as visualization tools, resulting in two distinct trajectories for her findings. Neither trajectory ended at findings: Following publication, the trajectory of an article continues as citations within related academic disciplines, standing in as markers for advances in representational capacity and method in computer science. But the trajectory of the representational tools followed another path, as they were used within the research of scientists. As we will see, Marie's rendering algorithms not only were used for dosage regiments but also came to be incorporated into the experimental systems of other kinds of scientific researchers, broadening the span of this material multidisciplinary link.

Mobile Code—Tracing the Multidisciplinary Link

The production of texture in Marie's work is only a piece of a visualization tool; in sum, a plethora of additional code is necessary to configure and render a full image. The particular visualization software features that I have traced in the first half of this chapter are not a whole rendering device, but merely individual algorithms among the many required. Much of this code will be borrowed, with modification, from previous applications.

Marie describes algorithms and tools as relatively mobile and adaptable: while many of her studies are tied to application *as well as* pursuing a research interest, each application does not "begin from scratch," she draws from past findings and on code from previous applications. With expert work, code can be transferred from one visualization application to another (Rolland and Monteiro 2002; Pollock and Williams 2008) just as many visualization specialists themselves move across domains.

In the case of Marie, the empirical studies that led to the creation of a tool for rendering images of medical dosage regimes later informed the production of a tool for visualization of vein topology. With what Marie described as "some tweaking of the code," her texture algorithms designed for representing dosage regimens were later utilized for producing an internal view of vein structure. Marie justifies this movement, or, more precisely, explains the logic permitting movement across disciplines, through the articulation of similarities between the data:

> Because we designed these algorithms by excluding extrinsic shape-markers [i.e., edges] in studying how the presence of texture might facilitate shape judgments across non-trivially structured interior regions where shape from contour information is not available, we understood the algorithm to be useful in cases where shape-edge is not available as a perceptual cue. (Publication)

The vein-structure data Marie modeled were taken from sensors placed within bodies. The sensors were thus "enveloped" in the body and could not provide data as to edges, or other extrinsic topology markers—this is the same case as with her original topology perception experiments that excluded markers such as shape. For Marie, these topological data generated comparable perceptual difficulties and thus their representation could be aided by using similar rendering techniques. Over the years she has built up a repertoire of "functionalities" alongside her research findings. Just as her findings build up over time, so too does her code base. Each finding in research inspires the writing of rendering code that operates according to the principles of that research.

The "functionality" of texture rendering described in the previous two sections was originally developed for permitting medical practitioners to visualize treatment regimens. Thereafter, Marie also worked closely with, for example, nanotechnologists in order to produce a similar effect with particle flows: maintaining a sense of particle movement without rendering cues such as edges. She described how selected portions of the program for visualization were transferred from one application site to another, with local adaptation traveling with Marie across received boundaries of disciplines and expertise. Marie recounted a familiarity with her own code and how this facilitated its incorporation into new applications. Her repertoire of code was amassed over time, each element informed by a particular empirical research project and application, but then "traveled" with her to new projects.

Here is a list of eight articles, in reverse chronological order, published by Marie and her coauthors:

1. Directional Enhancement in Texture-Based Vector Field Visualization
2. Illustrative Rendering Techniques for Visualization: Future of Visualization or Just Another Technique?

3. Visualization of Nanoparticle Formation in Turbulent Flows
4. Conveying Shape with Texture: An Experimental Investigation of the Impact of Texture Type on Shape Categorization Judgments
5. Line Direction Matters: An Argument for the Use of Principal Directions in 3D Line Drawings
6. Visualization Needs More Visual Design!
7. Art and Visualization: Oil and Water?
8. Strategies for Effectively Visualizing 3D Flow with Volume LIC

The ranges of her domain applications are quite broad. From the titles alone it is easy to see that her empirical applications in research are extraordinarily diverse, including the flow of nanoparticles (1,3), scheduling cancer radiation treatment (4,5) and mapping vein topology (4). Marie does display consistent interests, such as position papers on rendering with no direct reference to empirical applications (2,6,7) or theoretical discussions of how texture impacts shape perception (1,2,3,4,5,8).

These applications are from differing disciplines, departments and knowledge domains, tied together only by an underground stream of visualization technologies. While the fields of application are diverse, requiring expert domain science training, the particular visualization algorithms are not necessarily so. A flow representation for nanoparticles was informed by the same findings—and some of the same code—that was used for planning radiation exposure as part of cancer treatment.

Marie's algorithms may become incorporated in data visualization programs for domain scientists in innumerable academic and industrial fields. In using a data visualization produced by Marie's algorithms, her past work of capturing human perceptual properties vis-à-vis texture, of coding her findings or adapting previous code will be completely obscured from end users. These programs will produce detailed texture-maps that are intended to maximize particular topological features to the human eye. In each case of application to a new domain Marie describes ongoing interactions with domain scientists as she tailors her software for application. It is beyond the scope of this paper to explore these local modifications; instead I have focused on the backgrounded work that is common to all the applications (Pollock and Williams 2008). From the perspective of the trajectory I have explored in this chapter, each rendering can be taken as the end point in a long genealogy of technology design.

SINKING DEBATE INTO INFRASTRUCTURE

Such boundary crossing work is increasingly common to the sciences. However, in studies of multidisciplinarity it is usually approached as a matter of human-to-human collaborations. We have come to understand scientists to be the arbiters of representational validity within their domain. But with the redistribution of representational work described in this chapter, I ask, where do we locate authority over representation? Does a biologist have the technical expertise to interrogate the visual images produced by the software Marie designed for her lab?

Often the introduction of novel representational technique is a site of contestation (Bowker 1988; Bastide 1990; Daston and Galison 1992; Cambrosio, Jacobi et al. 1993; Gooday 1995; Edwards 1999; Golan 2004), but in this case the debates about technique occur in "another field"—in this case human-computer interface and data visualization. This separation of disciplinary research and representational work is significant for us as analysts and scholars of scientific visualization. Scholars in studies of scientific visualizations have come to agree that the knowledge of a science, and its consequences, are tied to representation devices and practice (Latour 1986; Suchman 1988; Lynch and Woolgar 1990; Daston and Galison 1992). This argument is occasionally extended by assuming that studying the visual elements of a science can be taken as a stand-in for practices in the science as a whole. For example, in his canonical article on visual languages in geology, historian Martin Rudwick outlines what has become a classic methodology for studies of scientific visualization:

> [V]isual means of communication necessarily imply the existence of a social community which tacitly accepts these rules and shares an understanding of these conventions. It is therefore worth studying the historical development of the visual language of geology . . . as a reflection of the growth of a self-conscious community of geological scientists. (Rudwick 1976: 150)

Rudwick takes visual conventions as a shorthand for the existence of an epistemic community with rules for evidentiary acceptance, of representational authorization and communicative coding and decoding. The visual language, and its historical evolution, is taken as a surrogate for the development of a self-identified scientific community of geologists.

Similarly, with carefully crafted historical detail, Cambrosio, Jacobi and Keating (1993) have outlined the case of the emergence of the visual conventions within Paul Ehrlich's work in immunology. Mirroring Rudwick's claim that representational technique and ontological entities co-emerge, they argue that Ehrlich's images are closely tied to the understanding of immunological entities themselves:

> The debate concerning the establishment of an immunological iconography appears to have been part of the constitution of immunological entities. . . . in order to record the existence and properties of entities, the development of a heuristic imagery was first needed, which would allow the work of inscription and representation to take place . . . the development of inscription or representation devices was cosubstantive

not only with the establishment of a given set of phenomena, but also with the constitution and definition of the nature and properties of the entities made responsible for those same phenomena. (Cambrosio, Jacobi et al. 1993)

In both Rudwick's and Cambrosio et al.'s work a tight configuration holds: What ontological entities exist *and* how to visually represent them were negotiated within the same disciplinary community of researchers. But with the redistribution of labor between information technologists and domain scientists described in the first half of this chapter, it becomes less clear that we as scholars of visualization may link the constitution and definition of the nature and properties of the entities with debates over representational convention. These debates may occur in only marginally overlapping social worlds.

Marie critiqued her own work, noting the deficiencies in interface design and statistical results: Did her experimental setup skew an understanding of shape perception? These criticisms were recorded in peer-reviewed publications, but they also independently informed the design of visualization tools for use in science and engineering. Ties between representational devices and epistemic commitments are weak (or nonexistent) in this configuration. Unearthing the controversies within HCI would require a kind of archaeological work unlikely for the "average user" of visualization software. Designers and users of the tool are separated by the gulf between publication and production.

—

In his study of the visual and statistical traditions within particle physics, Peter Galison (1997) demarcates three groups of practitioners within the community: theoreticians, experimentalists and instrument makers. He describes an "intercalation" of these groups: active collaborations that, at times, can be characterized as sharing epistemic commitments while at other times diverging. The various groups are primarily oriented to the activity within their own groups and may even be located in buildings, institutions or even countries that differ from the others. However, these diverging groups still interact productively in what Galison calls the "trading zone," where common language and modes of interaction have been established across professional specializations (Galison 1997). Here, Galison describes a *systematic disciplinary tie* between technicians, theoreticians and experiments. However, as I have described in this chapter, other configurations are also possible and increasingly likely.

We can roughly outline three kinds of relations between visualization experts and domain scientists. In the first case a group of domain scientists consistently work with their IT programmers through either a contractual relationship or shared research funding; the result is a tight interaction between social worlds. If the relations become systematic (or institutionalized) at the disciplinary level—as with Galison's physicists or the emerging fields of bioinformatics—a trading zone may emerge in which tropes for interaction and pidgins for communication develop. However, many researchers do not necessarily form such tight expert relations. For example, Star and Griesemer characterize how bird trappers and curators in a museum develop agreements about how to work together in gathering specimens. In this second configuration there is a loose familiarity across knowledge domains through collaborations that last years or decades but that are not based on a shared epistemic interest: Trappers do not "care" about taxonomy while still actively seeking to meet the informational requirements of curators, such as recording the location where a specimen was found (Star and Griesemer 1989). Susan Leigh Star describes this as "collaboration without consensus," a set of sustained relations that are quite different from the tight ties maintained by Galison's physicists.

Thirdly, the case I have explored with Marie is of a "light relationship," in which information technologists only loosely understand the domain but draw on combinations of novel, reused and prepackaged IT resources to provide tools *for* the domain. Visualization tools may be designed for one domain and then recompiled for another; here we have the movement of a black box, in which the inputs and outputs are only slightly adapted to the informational requirements of a domain. Marie consistently works with biology, but only in the broadest sense of the term. Her applications range from medical applications to the movement of subcellular particles. Software, or elements of code, is shifted across significantly heterogeneous epistemic communities. These are loosely framed collaborations of an informationally *adequate* nature that may be short-lived but thereafter materially solidified by relations through technologies of visualization.

In these cases, a correlation between community and the history of its rendering practice becomes murky at best and arguably altogether ruptured. Any approach to the studies of visualization in contemporary scientific practice should no longer *assume* continuity between representational and epistemic evolution. Continuity and rupture across disciplines must be investigated. The visualization conventions embodied in the applications that render data as image may be determined outside the confines of even the most broadly defined disciplinary arenas.

As scientific work becomes distributed in such novel configurations the analyst must be capable of tracing movements that, as the work of Marie shows, do not always occur on the "same disciplinary plane" as their use. This configuration also begs a much broader future research question: How do the automation of visualization and the distribution of representational work inform the co-emergence of representational form and ontological entities?

CONCLUSION: ELIDED CONFLICTS
AND REPRESENTATIONAL WORK

We have seen how visual representation conventions are imported from other disciplines or extrascientific institutions such as art; moreover, the techniques of representation have been slightly (sometimes radically) transformed. Marie's textures are neither woven nor drawn; they are algorithmically generated. Her tool is indirectly inspired, but in no sense equivalent to the loom that produced the original textures. They are *isomorphic renderings*, mirroring a visual appearance but not the method of production. Above and beyond this we have seen how research techniques such as experimental psychology come to inform the creation of those systems. Congealed as applications and technology, and used by scientists to visualize data in other fields, those findings sustain a multidisciplinary link that is difficult to trace.

Capturing the *arts and design* is a particular instance of visualization automation. I have met only a handful of researchers that explicitly define themselves as drawing technical inspiration from the arts. However, it *is* common for designers to study the "contexts of use" for technologies, to attempt to augment and support existing practices, and for digital design technologies to mirror their analog counterparts (Berg 1998). Rather than making a claim for generalizability, this research points to a social form: an organized activity that bridges between disparate media and links across disciplinary boundaries through tools.

The story I have told here is *not* a study in controversy, of "behind the scene battles informing the creation of a software function"; instead these are stories of an elided conflict, a possible tension left aside. Debate is always possible, but is made more difficult through the distributed organization of expert activity.

I have traced a backgrounded process for the creation of visualization tools. At the end of the process we do not have a representation, but rather a tool for building representation. It is a tool with particular properties—or affordances—but these properties cannot necessarily be deduced from the nature of the output representations. Most importantly, the use of the technology for visualizing data does not reveal the weaknesses in its design. In the case explored here, these weaknesses are documented in a stream of academic publications; it is possible to recover the flaws in rendering algorithms, but this requires a kind of archaeological work unlikely for most practicing scientists. It is *possible* for user-scientists to investigate the conditions of production for their visualization tools, but this is a challenging endeavor involving kinds of technical knowledges they are unlikely to have ready at hand.

Understanding the representational tendencies of a digital visualization technology requires understanding both their use in action (Amerine and Bilmes 1988; Suchman 1988; Winkler and Van Helden 1992; Prentice

2005) and the process by which they were produced; in this chapter I have traced the latter. Representational work is distributed between domain scientists analyzing data and information scientists who produce the means by which those data are analyzed, yet the work of both disciplinary groups never meets *in practice*; rather, they intersect at the product software.

REFERENCES

Amerine, R. and J. Bilmes (1988). "Following Instructions." *Human Studies* 11: 317–329.

Bastide, F. (1990). The iconography of scientific texts: principles of analysis. *Representation in Scientific Practice*. M. Lynch and S. Woolgar, 187–229. Cambridge, MA, MIT Press: 187–229.

Berg, M. (1998). "The Politics of Technology: On Bringing Social Theory into Technological Design." *Science, Technology & Human Values* 23: 456–490.

Bowker, G. C. (1988). Pictures from the subsoil, 1939. *Picturing Power: Visual Depiction and Social Relations*. G. Fyfe and J. Law, 221–254. London, Routledge: 221–254.

Cambrosio, A., A. Jacobi, et al. (1993). "Ehrlich's "Beautiful Pictures" and the Controversial Beginnings of Immunological Imagery." *Isis* 84: 662–669.

Daston, L. and P. L. Galison (1992). "The Image of Objectivity." *Representations* 40(Fall): 81–128.

Edwards, P. (1999). "Global Climate Science, Uncertainty and Politics: Data-Laden Models, Model-Filtered Data." *Science as Culture* 8(4): 437–472.

Galison, P. L. (1997). *Image and logic : a material culture of microphysics*. Chicago, University of Chicago Press.

Galison, P. L. (1999). Trading zone: Coordinating action and belief. *The science studies reader*. M. Biagioli, 137–160. New York, Routledge: 137–160.

Golan, T. (2004). "The Emergence of the Silent Witness: The Legal and Medical Reception of X-rays in the USA." *Social Studies of Science* 34(4): 469–499.

Gooday, G., J.N. (1995). The Morals of Energy Metering: Constructing and Deconstructing the Precision of the Victorian Electrical Engineer's Ammeter and Voltmeter. *The Values of Precision*. N. Wise, 239–282. Princeton, N.J., Princeton University Press: 239–282.

Hutchins, E. (1995). *Cognition in the Wild*. Massachusetts, MIT Press.

Jeffrey, P. (2003). "Smoothing the Waters: Observations on the Process of Cross-Disciplinary Research Collaboration." *Social Studies of Science* 33(4): 539–562.

Kay, L. (2000). *Who Wrote the Book of Life? A history of the genetic code*. Stanford, Standford University Press.

Latour, B. (1986). "Visualization and Cognition: Thinking with Eyes and Hands." *Knowledge and Society: Studies in the Sociology of Culture Past and Present* 6: 1–40.

Lynch, M. and S. Y. J. Edgerton (1988). Aesthetics and Digital Image Processing. *Picturing Power: VisualDepiction and Social Relations*. G. Fyfe and J. Law, 184–220. London, Routledge: 184–220.

Lynch, M. and S. Woolgar, Eds. (1990). *Representation in Scientific Practice*. Massachusetts, MIT Press.

Olson, G. M., A. Zimmerman, et al., Eds. (2008). *Scientific collaboration on the internet*. Cambridge, MIT Press.

Pollock, N. and R. Williams (2008). *Software and organizations: the biography of the enterprise-wide system or how SAP conquered the world*. New York, Routledge.

Prentice, R. (2005). "The Anatomy of a Surgical Simulation: The Mutual Articulation of Bodies in and through the Machine." *Social Studies of Science* 35(6): 837–866.

Rheinberger, H.-J. (2000). Cytoplasmic Particles: The Trajectory of a Scientific Object. *Biographies of Scientific Objects*. L. Daston, 270–294. Chicago, University of Chicago Press: 270–294.

Ribes, D. and G. C. Bowker (2009). "Between meaning and machine: learning to represent the knowledge of communities." *Information and Organization* 19(4): 199–217.

Ribes, D. and S. J. Jackson (2013). Data Bite Man: The Work of Sustaining a Long-Term Study. *"Raw Data" is an Oxymoron*. L. Gitelman, 147–166. Cambridge, MA, MIT Press: 147–166.

Ribes, D., S. J. Jackson, et al. (2012). "Artifacts that organize: Delegation in the distributed organization." *Information and Organization* 23(1): 1–14.

Rolland, K. and E. Monteiro (2002). "Balancing the Local and the Global in Infrastructural Information Systems." *The Information Society* 18(2): 87–100.

Rudwick, M. (1976). "The Emergence of a Visual Language for Geological Science, 1760–1840." *History of Science* 14: 149–195.

Star, S. L. and J. R. Griesemer (1989). "Institutional Ecology, "Translations," and Boundary Objects: Amateurs and Professionals in Berkeley's Museum of Vertebrate Zoology, 1907–39." *Social Studies of Science* 19: 387–420.

Suchman, L. (1988). Representing Practice in Cognitive Science. *Representation in Scientific Practice*. M. Lynch and S. Woolgar, 305–325. Cambridge, MIT Press: 305–325.

Suchman, L. (2007). *Human-Machine Reconfigurations: Plans and situated actions (2nd edition)*. Cambridge, Cambridge University Press.

Tufte, E. R. (1986). *The Visual Display of Quantitative Information*. Cheshire, Connecticut, Graphics Press.

Winkler, M. G. and A. Van Helden (1992). "Representing the Heavens: Galileo and Visual Astronomy." *Isis* 83: 195–217.

5 Making the Strange Familiar
Nanotechnology Images and Their Imagined Futures

Michael Lynch and Kathryn de Ridder-Vignone

INTRODUCTION

The idea that science and technology can be performed at the nanoscale usually is attributed to Richard Feynman's "There's plenty of room at the bottom" speech in 1959 at Caltech in Pasadena California (published in 1960).[1] In the speech Feynman imagined the possibility of writing the entire *Encyclopedia Britannica* on the head of a pin and reading it with an electron microscope.[2] He also envisioned the use of codes analogous to Morse code, but with "bits" made up of groupings of fifty atoms. Most famously, he envisioned the possibility of manufacturing with atoms to compose machines and electronic circuits. Feynman also made clear that manipulations of atoms would differ profoundly from manipulation of macroscale materials:

> When we get to the very, very small world—say circuits of seven atoms—we have a lot of new things that would happen that represent completely new opportunities for design. Atoms on a small scale behave like *nothing* on a large scale, for they satisfy the laws of quantum mechanics. So, as we go down and fiddle around with the atoms down there, we are working with different laws, and we can expect to do different things. We can manufacture in different ways. We can use, not just circuits, but some system involving the quantized energy levels, or the interactions of quantized spins, etc. (Feynman 1960, 36)

The promotion and eventual development of the transdisciplinary field of nanoscience/nanotechnology did not take off until decades later, when K. Eric Drexler and the Foresight Institute proposed the possibility of self-assembling machines (commonly dubbed "nanobots"—nanoscale robots). Aspects of Feynman's original proposal were preserved in Drexler's vision and became emblematic of nanoscience/nanotechnology ("nano" for short). These included proposals for research and engineering conducted at a scale of 1–100 nanometers (a nanometer is one billionth of a meter). Similar to what Feynman mentioned in his prophetic speech, Drexler and many other proponents of nanoscale research and technological development emphasized that the nascent field would need to contend with quantum mechanical forces that are radically different from macro- and microscale phenomena.

Drexler's idea of self-acting and self-assembling nanobots was contested, most notably by the late Richard Smalley of Rice University, who pursued a more self-consciously realistic vision of nano that was roughly consistent with Feynman's vision of a broad effort to etch, carve or assemble material products at an unprecedented level of scale. In many respects, nano followed the path of genetic and genomic research/engineering: It was funded by large government initiatives that strongly emphasized commercial potential; these initiatives drew together numerous scientific and engineering fields and developed new hybrid disciplines; there was no strict division between the scientific and technological aspects of research; and the field was promoted explicitly as "revolutionary" and "transformative" (sometimes with explicit use of Kuhnian idioms).

The short history of nano thus far includes a heavy dose of promises and fears. The futuristic scenarios envisioned at different stages of nano's history have been widely chronicled, as nano is often said to have left behind Drexler's visions of self-assembling and self-acting nanobots in favor of less dramatic possibilities for developing technologies such as medical tracers and pharmaceutical delivery vehicles, novel designs for fuel cells and molecular electronic junctions, resilient fabrics and clear sunscreens. Concerns about hazards also have moved away from visions of self-assembled nanobots proliferating beyond all human control, to more mundane concerns about the possible hazards of nanoparticles in air and water. Despite the general dampening down of the early hype, nano continues to be portrayed as a transformative field ushering in a distinctive mode of objectivity (Daston and Galison 2007, Ch. 7). Although we do not intend to suggest that there is nothing new under the sun (or at the very bottom, to borrow Feynman's idiom), we are impressed with the extent to which the emphasis on the twenty-first-century future of nano is conveyed through idioms and conventions that evoke much older, and even classic, modes of representation. We are also impressed by the diversity of things, images and modes of imagination that come into play in that field. Consequently, in this chapter, we shall inventory some of the different ways in which nano-images portray futures, often by making abundant use of classic artistic conventions and naturalistic portrayals of macroscale objects, surfaces and landscapes. Like their Renaissance counterparts (Edgerton 1976), the artisan-engineers of the twenty-first-century fashion artifacts that, at once but with different emphases, act as displays of craft as well as portrayals of things and scenes. And, like their predecessors, they compose images that formally embody conceptions of structure, thereby perfecting and idealizing the things they depict. More obviously than their predecessors, however, the artisan-engineers of the nanoscape produce *pretensional* images. This term is a modification of two terms (usually translated as "protention" and "pretention") that Edmund Husserl (1964, 1973) uses in discussions of the temporality of perception. Protention, the more familiar of the two terms, describes how an "act" of perception—or, in our conception, the sensible configuration of an image itself—reaches into the future by apprehending immediate details as part of an unfolding and coherent field. An auditory example is the recognition of a familiar musical piece from hearing the first few notes. The term "pretention" is related to Husserl's conception of intentionality, and describes how perception and consciousness tend toward (pre-tend) whole entities and expansive fields that are never fully perceived from an immediate vantage point or incorporated into an image. A different nuance is implied in English, when the term "pretension" is used to connote a suspect *claim* or imposture as well as a *tension* between a present moment and a possible future (Thomas 2004, 134). Although, from a phenomenological point of view, all images are protentional and pretentional, we believe that nano-images are perspicuous for the way they both claim and depict techno-scientific futures by drawing out imagined possibilities in the visible form of the objects and scenes they depict. And, as we shall see, their "pretensions" often have been subject to suspicion.

PROMOTIONAL IMAGES AND IMAGE GALLERIES

The promotion of nanotechnology reaches well beyond particular science and engineering disciplines and their immediate lines of patronage, and draws in broader publics. The prominence of such promotion is understandable, given the emphasis on transformative research, the attempt to foster a transdisciplinary convergence and hybridization of fields, and the concerted effort to attract and recombine ever-broader sources of university, commercial and government support. Images of objects, products and imagined possibilities are prominent at the interface between nano research and various actual and virtual publics. Beyond their use to depict phenomena and illustrate research articles, images—often multiple kinds of images—are exhibited in posters at conferences and in laboratory hallways; in one-page "nuggets" on laboratory websites that concisely present particular research projects and achievements; on journal and magazine covers; and in online image galleries. Such galleries, with their explicit reference to the public exhibition of art, are of particular interest for us in this chapter.

Unlike many "non-art" scientific images that are of interest to art historians and historians of science (Elkins 1995), many nano-images are produced and/or exhibited explicitly *as* art. A few of them verge on being "high" art destined to be preserved and revered for posterity. Some appear to be crude sketches and sculptures composed with unusual materials and tools; others are hyperrealistic computer graphic depictions of imaginary nanoscapes; and still others present what appear to be "found" shapes and structures that are framed, enhanced and colored in a way to provoke viewers' interest and curiosity. The use of the term "gallery" for collections of such images is an especially clear and explicit effort not only to suggest artistic connotations but also to *present* "scientific" images *as* art.

Galleries are, of course, best known as facilities that display works of art for sale and/or public appreciation. Online nano-image galleries differ from the archives and atlases that are of interest mainly to specialists. Such galleries are diverse, and open to all sorts of agendas, but the best known of them display and promote an individual's, company's, laboratory's or other organization's techno-scientific achievements and visions. It is common to find such images posted in the hallways of laboratory buildings, but the Internet is the most common home for image galleries. Because of the ease with which the digital images can be reproduced, amateur enthusiasts as well as researchers and research organizations compose galleries by collecting and exhibiting digital images and animations from one another's collections. Frequently, the same images are exhibited again and again on different sites, and a few have attained iconic status (Nerlich 2008, 275).

Galleries vary considerably, with some emphasizing spectacular visual effects and portraying imaginary (and even impossible) landscapes, while others present more prosaic scientific information. Although the captions in image galleries often say very little, either about the research that produced particular images or about the objects being depicted, the connections with science never entirely disappear. In many cases, at least a minimal science lesson is necessary for appreciating how such items are artful. In the absence of an understanding of the scale and technique of production, many of the images in nano-image galleries would appear to be crude and garishly colored sketches and abstractions.

The IBM STM Image Gallery

The IBM STM image gallery, originally created by Don Eigler and his group at IBM Almaden,[3] was one of the earliest and certainly most prominent of the online collections of nano-images. It is organized as a virtual gallery, with a floor plan of "rooms," such as "l'Hexagene" and "Blue Period," in which different genres of sci-art are displayed. Another well-known gallery, the Nanomedicine Art Gallery composed by Robert Freitas of the Foresight Institute, uses an evolutionary organization, as it presents various "historical species of nanobots" imagined by artists at different times. Other galleries are organized around the work of individual artists such as Chris Ewels. Still others, such as the Carl Zeiss Nano Image Contest, present winners of nano-image contests.

The IBM gallery displays the products of the scanning-tunneling microscope (STM), an instrument originally developed by scientists affiliated with that company. The STM is one of a family of probe microscopes that includes the atomic force microscopes and many other patented instruments

(Mody 2011). Probe microscopes operate in a fundamentally different way than traditional optical microscopes and transmission electron microscopes, which focus beams of incident or transmitted light or electrons to compose images. A scanning-tunneling microscope scans a surface by conducting a current of electrons from the surface to the tip of a probe, in a process known as quantum tunneling. The probe has an extremely thin tip that is chemically sharpened to a point that, in principle, is one atom thick. With current held constant, the device measures variations in the distance between the tip and the surface. Although there is nothing inherently "visual" about the resulting array of measurements, they can be (and typically are) converted into topographic images. Unlike an optical device, a probe microscope is often likened to a hand-held probe (such as the blind person's stick described by Michael Polanyi [1958, 64]), through which it is possible to "feel" the conformation of a surface, and even to inscribe figures in sand or push and pull small objects resting on the surface being probed.[4] However, the probe does not touch the surface, but instead conducts a flow of electrons from point to point across the surface it scans.[5] The measurements taken by the instrument are not easily translated into a topographic map of nanoscale hills and valleys traversed by the probe, because of the interaction between the probe and the electrical properties of the material it scans. The condition of the probe and of the scanned surface is unstable, and the current is subject to reversal when the probe picks up "junk" from the surface being scanned. When the tip becomes blunted by such "junk" so that it is less pointed than a promontory at the surface it is scanning, the flow of current can reverse and the surface of the tip momentarily becomes the "object" that is scanned.[6] Such hazards rarely are immediately visible, and are inferred from the output of runs and tests with the equipment.

It is ironic, perhaps, that visual images have been so prominent during the short history of nanotechnology. The nanoscale is positioned well below the minimum wavelengths of visible light. As Feynman mentioned in his famous speech, and many others have reiterated since then, taken-for-granted features of macroscale objects and scenes, such as color, solidity and discrete boundaries, have no place in the nanoscape. Nevertheless, researchers and image processing specialists go to great lengths to compose images that make abundant use of familiar visual forms and macroscale qualities.

According to Jochen Hennig (2005), starting in the late 1980s STM measurements were shaped into pictures that appeared to visualize individual atoms. Hennig details how measurements of voltage made while the tip of the instrument scanned a surface, with current held constant, were combined and composed into pictures, with individual atoms conventionally depicted as color-coded peaks. This convention of depicting atoms as rounded conical peaks is prominent in the IBM gallery. Hennig observes that the visualization of such peaks fulfilled expectations that atoms should appear as elevations of the topography of a surface: "No longer is the path of the tip transformed into an image. The individual constituent parts of the measurements have been put together in a manner that suggests an apparent reproduction of the investigated object" (Hennig 2005, 156). Such "atoms" were thus depicted in a way that conformed to conventional macroscale models and depictions of atoms as discrete things arrayed in patterns.

Such images depicted objects with clearly bounded edges and color variations marking out distinct entities and regions. Frequently, these depictions made conspicuous use of classic artistic conventions: vanishing points, chiaroscuro and other aspects of linear perspective that suggest illuminated landscapes viewed from a fixed perspective.[7]

The IBM gallery makes use of such conventions for depicting atoms, most obviously with the much-reproduced image of the letters "IBM" composed with a series of xenon atoms depicted with false-color as iridescent blue peaks arranged on a dark background (identified as a nickel surface).[8] Not only is the STM used to arrange the configuration it represents, but also the image in this case conspicuously signifies the corporate patron of the research—one may be reminded of Galileo's naming of the "Medici stars" (now regarded as the moons of Jupiter) after his patron (Biagioli 1990), though Galileo lacked the ability directly to write the name "Medici" in the night sky.

Another set of images, in a virtual room of the IBM STM gallery punningly named "The Corral Reef," depicts a series of "quantum corrals." The particular corral image in Figure 5.1 has gained widespread circulation in scientific publications (Crommie, Lutz and Eigler 1993) both as a research object and as cover art, as well as in many popular websites and publications. The iron atoms are represented by the color-coded conic peaks. Like the IBM logo-image, the corrals are assembled by moving the atoms with the STM tip, in this case to form a ring.[9] However, unlike the IBM image, the composition *itself* becomes an instrument for generating and investigating a physical phenomenon: a pattern of electron waves that are depicted within the "corral."

Without discounting the novelty of the quantum corral, we can note that the depiction in Figure 5.1 deploys some highly conventional artistic devices. Computer graphics were used to produce the color composition, and to construct the visible shapes, discrete surfaces and light-and-shadow effects. The landscape can be likened to a surreal volcanic caldera of bubbling lava (a highly symmetrical and smoothed rim and pool of lava in this case), but these (sur)reality effects are a consequence of the artist's deployment of classic conventions of linear perspective to suggest an imaginary observer hovering above and close to the surface of the quantum corral. As Chris Toumey (2009, 151) expresses in his analysis of this ubiquitous image, "Together the angle and the scale gave the sense of peeking over a fence to see the electron wave within—the sense that the viewer could almost touch the fence of atoms (as if atoms were solid objects)." The surface is discretely bounded, apparently illuminated by sources of light (the

Figure 5.1 "Quantum Corral" (1993), Part of *The Corral Reef* part of IBM's STM Image gallery. Image originally created by IBM Corporation. Creator: D. Eigler. Available at: http://www.almaden.ibm.com/vis/stm/atomo.html.

shadowing on the blue peaks implicates a source of illumination to the left, while the molten pool appears to be illuminated from a different source to the right and front of the viewer).

TRUST, DISTORTION AND MAKING THE STRANGE FAMILIAR

Widely circulated images, such as the "corral" reproduced in Figure 5.1, have been criticized for distorting the STM data and misleading viewers into believing that they are viewing colorful, three-dimensional fields illuminated with visible light. Chris Robinson (2004, 167), for example, points out that creators of such popularized images often "take extensive license in making an image that is often more visually interesting than scientifically informative." He expresses concern that such digital manipulation will create "a loss of trust in the visual truth of photographs" (ibid., 169). Philosopher Joseph Pitt also objects to the way "these computer-generated and enhanced pictures suggest that the world is at rock bottom a simple place. It can be pictured as individual atoms resting on stable fields that we can manipulate at will, twirl them, enlarge and narrow them, put them to music, make them dance, when in fact nothing of the kind is the case. The world at the nano and quantum mechanical level is . . . buzzing, shifting, constantly in motion in non-linear and non-classical causal fashion" (Pitt 2005, 31; see also Ottino 2003).[10]

Robinson, Pitt and other critics of popularized nano images recognize that *all* images of nanoscale phenomena—including those published in technical articles and used analytically by practitioners—must make use of visual analogies and conventions to make such phenomena sensible and intelligible. They also recognize that literary and pictorial devices that serve to make strange (nonvisual, theoretical, abstract) things familiar (visible, intelligible, describable) are far from unique to nano. As we understand their criticisms, however, Robinson and Pitt are not simply complaining that popularized nano images "distort reality." Instead, they are pointing to a missed pedagogical opportunity. If, as Feynman emphasized, and many others have repeated in the decades since his visionary speech, nanoscale entities in quantum fields are *nothing like* macroscale forms and Newtonian forces, then popularized images that depict them as though they were familiar kinds of objects, illuminated with light and viewed from a particular vantage point, miss the opportunity to challenge conventional, naturalistic ontology.

Making the Strange Familiar

The theme of "making the strange familiar" is especially obvious when we examine nano-images prepared for online digital galleries and databases. This theme is familiar from previous studies of nanoscale images. Nerlich (2008, 289), for example, speaks of features of nano-images that function "to make the unfamiliar familiar" (also see Nordmann 2004; Milburn 2008). The particular turn of phrase we are using is an inversion of Garfinkel's (1967) theme of *making the familiar strange* through methodological interventions designed to create bewilderment, and thus to expose and elucidate the fragile underpinnings of taken-for-granted, everyday activities. In the case of popularized nano-images, properties of objects and fields that are said to behave in ways that break with familiar conceptions of reality and objectivity nonetheless are depicted as familiar sorts of things in recognizable landscapes. Now, of course, few of us are accustomed to seeing atoms arranged in "corrals." However, we are likely to be familiar with conventional pictorial uses of lighting, shading, color and surface, as well as with textbook depictions of atoms and molecules. The conventions differ from case to case, but we can identify some typical features of particular kinds of nano-images that appear in galleries and related forums for exhibiting and promoting nanoscale "art." There also are differences in the analogies that come into play in different contexts.

AN INVENTORY OF NANO-IMAGES

Like scientific images in other fields, nano-images take different forms in different contexts. There are striking differences in how particular galleries highlight the relevant relations of production, the actual and possible uses, and the public reception of nano-images.

An inventory of such images can convey an appreciation of these differences. Such images do not sort themselves out into a single array of categories, or even a single spectrum along which different images can be placed.[11] Ours begins with the most empirical and arguably artless renderings, followed by more schematized portrayals that perform quite different representational functions, often in combination. Some conspicuously exhibit technological achievements at an unprecedented level of scale; others appear selected and framed to evoke the appearance of macroscale *objects and landscapes*; still others exhibit structural form with *idealized models*, often depicted in "real" environments; and, finally, some evoke science fiction imagery of *fantastic voyages* through hyperreal nanoscapes. Each type of image involves distinct modes of presentation and distinct combinations of imagination and realism. It is not as though the most "empirical" are the least imaginative, or that the most imaginative are the least realistic. As in the history of art (and no less in the history of science) there is no single standard of realism, and no single modality of imagination. Instead, we find complex juxtapositions and gradations, sometimes within the frame of a single image.[12] And, as we shall argue, the most *ostentatiously* artful, imaginative and even fantastic images often are the least effective at provoking an imagination of objects that, in Feynman's terms, "behave like *nothing* on a large scale."

Empirical Renderings

Comparatively few explicitly empirical images appear in nano-image galleries, although they are common in technical publications and laboratory website "nuggets" (compact web pages presenting a laboratory's wares). By "empirical" we refer to the aesthetic quality of unadorned grayscale or monochrome renderings of data recorded from instruments, which are used to depict phenomena of interest to specialist researchers (also see Ruivenkamp and Rip, 2010, 17). Such images present the kind of unadorned "warts and all" appearance that Daston and Galison (1992, 82) identify with the epistemic and moral quality of mechanical objectivity. The few such images that can be found in image galleries often are pictures of "nanotubes" and "buckyballs" (or "fullerenes," named after Buckminster Fuller's geodesic domes), both of which are described as highly regular molecules composed solely of carbon atoms.[13] In STM, transmission electron microscope and scanning electron microscope images, buckyballs and nanotubes tend to show up as blurry, irregularly shaped forms that often lack interior detail. When depicted as models, they are perfectly regular, with nanotubes resembling cylindrical rolls of chicken wire, and bucky-balls taking the form of soccer balls composed of five- and six-sided carbon rings. And, as we discuss below, these models are often placed in pictorial frames that endow them with vivid macroscale characteristics.

Nanocraft: Displays of Technical Virtuosity

Many nano-images displayed in image galleries appear to be photos of mac-roscopic objects. Viewed out of context, they appear to depict oddly con-formed flowers or mushrooms, crudely sculpted figures and artifacts, and rough etchings or cartoonish sketches of recognizable figures and faces.[14] However, a minimal tutorial about the scale of the depicted objects alerts the viewer to appreciate the "art" they embody. It is necessary to take into account that they were produced at an incredibly small scale with materials that differ profoundly from what a macroscale artist has at hand. Marshall McLuhan's (1964) slogan "the medium is the message" surely applies to these images, though perhaps more literally than he intended. The images are demonstrations of what the artist can *do* with equipment, such as an STM, working in a novel medium. Compared with illusionist art (Gom-brich 1960), the point of these images is not to create simulacra that hide the craft that composed them. Instead, they are displays of virtuosity that draw attention to the artist's "haptic" manipulation of a material medium (Daston and Galison 2007, 383).[15]

The IBM STM Image Gallery includes several such images (including the famous IBM composition with xenon atoms) in its "Atomalism" room. In the absence of any indication of scale and composition, some of these figures (for example, the "Carbon Monoxide Man" [http://www.almaden.ibm.com/vis/stm/atomo.html]) appear to be stick figures composed of metallic dots. Once it is appreciated how small these depicted figures are, and that they are composed of molecules or atoms arranged on a platinum, gold or graphite surface, they become displays of extraordinary technical virtuos-ity. They show *that* materials were manipulated through novel, skillful and imaginative uses of probe microscopy to compose recognizable figures.[16]

Many nano-images in this category (or, in many image galleries, images of larger-than-nano entities imaged with scanning electron microscopes) are pictures or micrographs of three-dimensional objects. These include chemi-cal structures and mechanical sculptures produced through "bottom up" atomic and chemical assembly or "top down" manipulation or sculpting with external instruments. An example that nicely exhibits the playfulness associated with many of these nano art objects is Takahashi Kaito's "small toilet," a nano sculpture, etched roughly in the form of a toilet, which won the "Most Bizarre prize" at the 49th International Conference on Electron, Ion and Photon Beam Technology and Nanofabrication Bizarre/

Beautiful Micrograph Contest (www.nanowerk.com/news/newsid=2840. php, accessed December 22, 2012). This toilet is not a "ready made," like Marcel Duchamp's notorious urinal displaced from its usual facility to the gallery; instead, the image testifies to the achievement of making the object. The photographic documentation of the fact *that* a sculpture was created at the nanoscale is the key to its appreciation; the sculpted material remains essentially out of sight and out of reach.

Unlike NanoToilet, some nano-sculptures are said to have functional properties, analogous to those of the things depicted, though working with very different physical principles. An example is a nano guitar with "strings," attributed to a doctoral student, Dustin Carr, at the Cornell University Nanofabrication Facility ("New Cornell" 2003). Accounts of this "guitar" allege that it can be activated with an electron beam. The aesthetic quality of the "music" played on the nanoguitar is far less significant than the fact *that* something analogous to music can be "played" with the thing. Another example of such semifunctional nanotechnology, which we discuss later, is "nanocars" with "wheels" that supposedly turn on their axles.

Self-Assembled Objects and Landscapes

Nano researchers use a vernacular distinction between "top down" and "bottom up" relations to the objects they compose. "Top down" refers to using instruments to carve, etch or manufacture the nano-object of interest with a tool, such as the STM tip, whereas "bottom up" refers to "self-assembly." The latter refers to atomic or molecular crystals, surface anomalies and other complex structures that form when researchers place molecular ingredients in solution, so that they will interact to form particular structures. Self-assembly is sometimes predictable, but it can also produce surprising and fortuitous features.

Gallery displays often exploit and highlight the resemblance between nanoscale (interpreted loosely, as they often are orders of magnitude larger) entities and macroscopic objects. Abundant examples can be found in image galleries of pictures that appear to depict clusters of "grapes" or bouquets of "flowers."[17] The placement of these pictures in nano-image galleries, and the brief captions that accompany them are sufficient to alert viewers that these objects have nothing to do with grapes and flowers, but instead are crystalline formations or clusters of nanoparticles found at a scale millions of times smaller than the familiar macroscale objects they resemble. The art is not unlike that of a nature photographer who selects and frames a subject to produce striking, sometimes abstract effects. Though scale is crucial, and is often indicated within the frame of such images, appreciating these images often requires minimal conceptual preparation.

Other images compose and reframe nanoscale constructions to take the form of recognizable landscapes. An example in the IBM STM Image Gallery is a composition entitled "A Copper Perspective" (Crommie, Lutz and

Eigler 1993). The image is accompanied by a caption that likens it to a formal Japanese rock garden, though the color scheme and layout seem equally reminiscent of a Southwest US (or even Martian) landscape. A common digital imaging technique used for these landscapes is to display the surface as a perspectival vista, rather than as a two-dimensional "map" viewed from directly above. In some cases such landscape views highlight significant three-dimensional features, but in popular galleries they appear to be chosen for spectacular effect.

Idealized Images/Models

Computer images and animations of nanoscale objects characteristically are "smoothed"—as though airbrushed to convey a sense of clean, uniform structures shown in clear space and free from contamination (Curtis 2007; Goodsell 2009, 54). Robinson (2004) identifies a broad class of such model-depictions as "schematics"—images that show abstract or idealized structures. However, what particularly interests us about many such images is that they endow schematic forms with details that suggest concrete materiality. As we noted earlier when discussing the quantum corral in Figure 5.1, conventions of linear perspective and chiaroscuro are employed to lend these images hyperrealistic qualities. These conventions also are employed in depictions of abstract molecular models of "nanotubes" and "buckyballs." Many examples of this kind of depiction can be found in artist/scientist Chris Ewels's "Nanotechnology Image Gallery" (http://www.ewels.info/img/science/index.html, accessed December 22, 2012). Ewels's images, like many others in galleries and publications, deploy the conventions of the space-filling and ball-and-stick models (Goodsell 2009, 56) that were used decades ago as hand-tools in chemistry and molecular biology to reconstruct and display chemical structures that are never fully visible through any single imaging technology (Francoeur 1997). Ewels depicts phenomena that appear to be composed of billiard ball "atoms" joined by cylindrical bonds, or of the inflated rubbery components of space-filling models. However, he depicts these idealized models, with their regular, repeating structures, and absence of any flaws or anomalies, as though they were macroscale (or, in some cases, megascale) objects depicted in voluminous space. An example is his composition "Floating Fullerines," which depicts a buoyant and rubbery space-filling model floating on a watery surface (http://web.archive.org/web/20071225194447/www.ewels.info/img/science/gallery/index.html, accessed December 22, 2012).

Ewels's images of carbon nanotubes also show highly regular structural details, unlike images composed from STM data in which depicted nanotubes appear to have an irregular shape and undifferentiated surface. Ruivenkamp and Rip (2010, 18) point out that an "artist's impression" composed in the manner of Ewels's images appeals to popular audiences by building macroscale form and perspective into the composition. However, this type of rendering also reveals inferred chemical structures that are not concretely visible in "empirical" micrographs.

In some cases, the depicted models show more than chemical structures; they also can suggest atomic or molecular mechanisms analogous to those in macroscale machines, such as Foresight Institute founder K. Eric Drexler's model of a "molecular differential gear" (www.imm.org, accessed December 22, 2012). The component atoms are depicted as color-coded spheres, packed together in an assembly that resembles a macroscopic gearing mechanism.

Such imagined mechanisms were emblematic of Drexler's vision of molecular manufacturing, which became central to his debate with the late Richard Smalley that ran in the pages of *Scientific American* and *Chemical & Engineering News* from 2001 through 2003 (see Kaplan and Radin 2011). Smalley and other proponents of nanotechnology made concerted efforts to encourage what they believed to be more sober visions of the future—sometimes characterized facetiously as a move toward "denanobotization" (Lösch 2008, 125). As noted earlier, the more mundane and less controversial image of nano that is now offered downplays early visions of self-assembling machines. Nevertheless, similar color-coded assemblies of molecular mechanisms are presented as structural renderings of synthetic molecular machines such as rotaxanes. These depictions identify the symbols of chemical formulae with familiar macroscopic mechanisms (Ruivenkamp and Rip 2010, 2014; Browne and Feringa 2006; Schummer 2006), as well as biological mechanisms such as bacterial flagella (Myers 2008, 2014).[18] Interestingly, some of the most cutting-edge research on such mechanisms is taking place at Rice University's Richard E. Smalley's Institute for Nanoscale Science, where "nanocars" consisting of buckyball "wheels" bonded to alkyne "axles" are being designed. The nanocars are said to "roll" on a surface when heated or pulled by the tip of a probe microscope (Figure 5.2).

Figure 5.2 is particularly interesting for the way it juxtaposes a schematic model (left) of a nanocar with a monochrome STM image of an array of such "vehicles" arrayed on a gold surface (right). The model vividly uses the conventions of color-coded spheres to depict a nanocar rolling over a clean, regular, and structurally precise surface of uniform gold atoms. The model is accentuated with artistic conventions of linear perspective and chiaroscuro, with shadow effects applying even to the diagrammatic arrow. The image positions the viewer in front and to the side of a clearly distinguished vehicle with four wheels joined by axles. In contrast, the STM rendering on the right identifies the "nanocars" as the small, lightly hued clusters, viewed from above as though they were an army of rovers on Martian terrain. In many of them, four "wheels" can be distinguished, but the rotary motion cannot be inferred from a single image—it was inferred through difficult and exacting comparisons of a series of STM images as

Figure 5.2   Nanocar models and STM image. Reproduced from Yasuhiro Shirai, Andrew J. Osgood, Yuming Zhao, Kevin F. Kelly and James M. Tour. 2005. "Directional Control in Thermally Driven Single-Molecule Nanocars." *Nano Letters 5* (11): 2330–2334. Copyright 2005, American Chemical Society. Reprinted with permission from American Chemical Society.

well as other forms of evidence. The model synthetically depicts a theoretical movement, while the STM image provides it with an empirical complement that lends an accent of reality to what has been extrapolated from a series of microscopic frames (see Lynch 1988 for such pairings in earlier forms of microscopy).

Fantastic Voyages

Some of the fantastic nano machines and scenarios that proliferated during the early "visionary" phase of the nanotechnology movement do not resemble molecular models, but instead resemble macroscopic machines. In a manner reminiscent of the 1966 film *Fantastic Voyage*, such machines are shown traveling through arteries, detecting sources of trouble and repairing the damage. Red blood cells are common reference objects in these depictions, and are used for establishing scale and place. The depicted vehicles appear to be made of hard stuff—metals, plastics—and their designs are reminiscent of fictional submarines, spaceships and robots. These images are freely drawn, though in hyperrealistic fashion. They sometimes have a "retro" look to them, reminiscent of early to mid-twentieth-century science fiction paintings of Martian landscapes and fantastic voyages of discovery—what Nerlich (2005) calls "vehicular utopias." For example, the "historical species of nanobots" in the Nanomedicine Art Gallery, representative of the 1980s and early 1990s, transpose spaceships, submarines and robots to the appropriate scale to conduct nano-medical missions.[19]

These depictions and animations completely disregard the idea that the quantum mechanical forces that are said to operate at the nanoscale are essentially unlike what is found at higher levels of scale. The machines in these images also tend to be several magnitudes larger in scale than the typical range ascribed to nano. Though evoking novelty and adventure, these images are strikingly conservative in their deployment of "traditional imagery" using a familiar style of artwork for depicting undersea and outer space voyages (Nordmann 2004; Nerlich 2005). A typical picture shows a vehicle traveling through a capillary filled with a completely transparent plasma medium surrounding textured and well-illuminated blood cells. Such fantastic depictions are characteristic of an early genre of nano-art, which is on the wane in the era of "de-nanobotization," though "fantastic voyage" imagery and analogies continue to inspire research in, and promotion of, nanotechnology (see, for example, Sengupta, Ibele and Sen 2012).

DISCUSSION AND CONCLUSION

Each of the depictions or models we discussed is "realistic" in one or another sense of the word, and each also calls into play one or more distinctive modes of imagination. Empirical renderings are realistic in the sense that they retain the gritty appearance of unadorned, monochrome or black-and-white, micrographic images. These images are highly processed and worked over—they are not simply delivered mechanically by instruments. However, they are presented as exhibits of microscopic data, rather than structural models or freely drawn or sculpted objects and landscapes. The images are problematic in the way the instrumentation limits and intrudes upon what it makes visible, so that the resulting images place a heavy burden on the viewer to imagine what, if anything, the image shows.

Images that display technical virtuosity engage artistic imagination through demonstrations of how exotic materials can be worked, and even made functional, at an unprecedented level of scale. While they often exhibit a playful sensibility, they demonstrate what *can be made* at the nanoscale—in other words, they exhibit the combination of practical realism and imagination that patent law dubs "enablement" (the requirement that a claimed invention must be realizable), and like a patent claim they require *imagination*, and not concrete demonstration, that they *can possibly* be made, and made useful. They compare interestingly with the early "species" of nanobots depicted in the Foresight Institute's gallery. Few proponents of nanotechnology today would say that Eric Drexler's visionary nanobots *can* be made real, at least not as originally envisioned, whereas nobody doubts that it is possible to spell out "IBM" by manipulating xenon atoms with an STM, or to create a "nanoguitar" (though the "music" it plays is another matter). There are doubts that the accomplishment of such technical feats indicates the most promising way forward for nanotechnology, but there is no serious question that it is possible to create the figures, inscriptions and sculptures depicted in these images.

Images of self-assembled objects and landscapes display "found" rather than "made" forms, as though conveyed through a kind of nature photography. The distinction is problematic, in so far as the "found" objects are "made" through the arts of chemistry and materials science, but unlike nano-inscriptions and sculptures, the particular forms selected for an image often have fortuitous features. Framing the image and naming the apparent object or scene are crucial spurs to the imagination. To appreciate these images it is necessary to know that the things they depict are not what they appear to be: they are not mushrooms, flowers or desert landscapes, but nano-structures created by chemical and mechanical means. In many cases, they also exhibit the surprise of encountering unexpected empirical formations (such as crystalline structures) that were not explicitly planned for or designed. Such fortuitous structures are the stuff of discovery as well as invention.

Images constructed along the lines of molecular models fill the breach between concrete image and imagined object with purified theoretical renderings. They are realist in a "Galilean" sense (Husserl 1970, 23–59)—that is, they invest the accent of reality not in the sensual qualities of the perceived (or imaged) thing, but in its underlying geometrical structure. At the same time, however, many renderings of models in image galleries enlist the techniques of artistic illusionism to create sensory reality effects in which chemical structures are depicted as colorful, illuminated and perspectival *things*. In this sense they straddle two modes of "truth to nature": apparent fidelity to a mathematized nature that cuts through the illusions of the senses; and fidelity to the appearance of an illuminated, voluminous, perceived space depicted by exploiting those very same illusions. For objects that are presumed to exist, and for which a given model (for example, of buckyballs or nanotubes) is deemed structurally accurate, the model may provide more detailed information about structural properties than any empirical rendering can possibly provide. Models that suggest possible macroscopic entities and mechanisms such as nanocars and Drexlerian machines enable viewers to construe chemical structures "as if they were ordinary objects," and to "imagine these molecules performing all kinds of functions" that the macro-objects do (Schummer 2006, 61). Even if such extrapolations are "working conceptual hallucinations" (Gilbert and Mulkay 1984, Ch. 7), there is no solid unhallucinated image with which to compare them.

As noted earlier, the "fantastic voyage" images explicitly call upon the styles and scenarios of science fiction. They present fantasy scenarios in the form of pictorial illustrations of machines that look like submarines or spaceships traveling through colorful, well-lit environments. Ironically, while such images are ostentatiously imaginative in the way they depict futuristic scenarios, in another sense they are the *least* imaginative and most "epistemologically nostalgic" (Lynch 1991, 72) of the types of image we discussed. Their heavy use of macroscale form and pictorial realism does not challenge viewers to imagine the distinctive modes of existence and means of visualization that are said to operate at the nanoscale. Instead, such images depict the strangeness of nanoscale objects and environments in an all-too-familiar way. Indeed, though still ubiquitous in many galleries, images of nanobots navigating through fields of blood cells are fading from salience; perhaps by now they have succeeded in making the strange *boring*. As noted earlier, all of the images in online nano-image galleries—and by extension, all images of nanoscale "things"—necessarily trade upon conventions for making the strange familiar. This does not mean that they are all liable to charges of distortion, since such charges beg the question, "Distortion of what?" Although we share concerns about the extent to which image galleries inform or mislead the public, we also are aware that galleries are quite diverse, are directed at different audiences, and reflect different purposes. And, given their accessibility, they are subject to endless reappropriation and recombination. The more informative images do not avoid, let alone hide, manipulation, but instead display *that* and to some extent *how* they were manipulated. In addition to depicting different futures imagined during different eras in the short history of nanotechnology, image galleries also serve different purposes, such as engaging broad publics, or calling the attention of potential funders to cutting-edge research.

It is frequently noted that nanotechnology is a future-oriented movement, and nanoimages express such orientation by depicting future possibilities and fantasies as part of a campaign to promote the nascent field and build public support for it (Milburn 2002; Lösch 2008). Worries about generating unrealistic hopes and fears shadow the promises, pretenses and pretending expressed in popularized writings and images associated with the movement. Earlier, we suggested that these images are pretensional (though not necessarily pretentious), as they concretely depict what might (arguably, imaginably) be constructed in a near or distant future. They are not *performative* in the sense of serving immediately to effectuate what they depict, since in some cases they portray what is (arguably) impossible to construct and in other cases the possible performance faces immense challenge. Moreover, they do not simply engage the imagination by envisioning futures. Although such futurism has been a prominent, and even characteristic, feature of nanoimages, many such images engage the viewer's imagination by evoking recognizable objects, scenes, forms and artistic conventions. Moreover, they do so not only by showing what *might* become a reality in the future but also by artfully displacing familiar macroscale forms and relations into the nano domain. Many popular depictions present structural abstractions, represented as space-filling or ball-and-stick molecular models, *as* material objects in concretely witnessed, lived-in, perspectival landscapes. The pretension thus reaches into space as well as time, to domesticate the blooming, buzzing confusion of quantum-mechanical space, and to place it safely on screens manipulated by keyboards in the safety of home and office.

ACKNOWLEDGMENTS

The research for this paper was funded in part by a 2008–2009 Scholars Award from the National Science Foundation, Science and Society Directorate, SES-0822757, entitled "Visualization at the Nanoscale: The Uses of Images in the Production and Promotion of Nanoscience and Nanotechnology."

NOTES

1. A transcript of the speech is available at: http://www.zyvex.com/nanotech/feynman.html (accessed March 14, 2012). More interestingly, a transcription of a small portion of the speech was inscribed at the nanoscale by means of the technique of "dip pen lithography," and can be viewed online at: http://www.boiledbeans.net/2009/09/06/squint-your-eyes-a-lot/ (accessed March 14, 2012). Toumey (2008) argues that the significance of Feynman's speech is largely a post-facto construction in light of much later developments.
2. Interestingly, as we write, the *Encyclopedia Britannica* has announced that it no longer will continue its print editions, but will be entirely online (see Bosman 2012).
3. The IBM STM Gallery is available at: http://www.almaden.ibm.com/vis/stm/gallery.html (accessed March 30, 2012).
4. Lorraine Daston and Peter Galison (2007) ascribe a novel mode of objectivity to the "haptic" mode of apprehension through which a probe microscopist simultaneously manipulates and visualizes the configuration of a surface.
5. An illustrated explanation of the STM's operation is available in the "Lobby" of the IBM STM image gallery: http://www.almaden.ibm.com/vis/stm/lobby.html (accessed March 30, 2012).
6. Kevin Kelly, Rice University, personal communication, February 2009.
7. Chris Toumey (2009) also recognizes the way classic linear perspective is deployed in images such as the quantum corral, but he also emphasizes that such images invite viewers to deploy multiple perspectives in a way that he likens to cubism.
8. The IBM image and similar figures are available in the "Atomalism" room of the IBM STM gallery: http://www.almaden.ibm.com/vis/stm/atomo.html (accessed March 30, 2012).
9. In a video, Eigler explains how he manipulates atoms, and likens his craft to a child working with Lego blocks. Available at: http://www.physorg.com/news173344987.html (accessed March 30, 2012).
10. See Emma Frow's analysis of particularly acute concerns with "Photoshopping" in cell biology (Frow 2012, 2014).
11. A version of this inventory is in de Ridder-Vignone and Lynch (2012). For related typologies and analyses of nano-images, see Robinson (2004, 2012); Nerlich (2008); Toumey (2009); and Ruivenkamp and Rip (2010, 2014).

12. Because of the limit on the number of illustrations we are permitted to reproduce in this chapter, when discussing particular examples we shall provide references to online sites. In the event that particular images are no longer available at the URLs we cite, in most cases they should be available at other sites.

13. We thank Nils Petersen of the University of Alberta for correcting an earlier version of this and other technical points in this chapter. The Wikipedia entry for "carbon nanotube" (http://en.wikipedia.org/wiki/Carbon_nanotube) includes numerous illustrations, most of which are diagrams or models. However, several of the images are identified as transmission electron, scanning electron or scanning-tunneling microscope images. It is these types of images that we are referring to as "empirical renderings."

14. For a collection of such sculptures, see: http://gadgetzz.com/2011/08/23/nano-art-why-would-you-make-a-10nm-toilet/ (accessed March 30, 2012).

15. Typically, these images show *that* a technical feat was accomplished, but without showing *just how* it was made. The recognizable form of the object stands proxy for the accomplishment.

16. In an analysis of televised scientific demonstrations, H. M. Collins (1988) distinguishes between experiments that require expert interpretation to understand and criticize how the experiment was done, and public "displays of virtuosity" designed to convince an audience *that* a credible scientific test was done. We use the expression "displays of virtuosity" to identify a primary function of many images, which is to display craft and skill (whether to other practitioners or to lay audiences) instead of, or in addition to, depicting an intrinsically interesting object or scene. These two functions are far from incompatible, as displays of virtuosity can support theoretical analyses with demonstrations of technical expertise (Mody and Lynch 2010).

17. See, for example, the winning entries in a contest sponsored by the 52nd International Conference on Electron, Ion and Photon Beam Technology and Nanofabrication. Available at: http://www.zyvexlabs.com/EIPBNuG/EIPBN2008/2008.html (accessed December 22, 2012).

18. Myers (2008) points out that proponents of "Intelligent Design" theories (arguably associated with religious creationism) use modeled portrayals of the protein mechanism that powers the flagellum. Depicted at the base of the flagellum is an arrangement of molecules drawn in the manner of Drexler's differential gear, suggesting a machine and implying a "designer."

19. Available at: http://www.foresight.org/Nanomedicine/Gallery/Species/HistorGeneral.html (accessed December 22, 2012).

REFERENCES

Biagioli, Mario. 1990. "Galileo the Emblem Maker." *Isis* 81 (2): 230–258.

Bosman, Julie. 2012. "After 244 Years Encyclopedia Britannica Stops the Presses." *New York Times*, March 13.

Browne, W. R. and B. L. Feringa. 2006. "Making Molecular Machines Work." *Nature Nanotechnology* 1 (1): 25–35.

Collins, H. M. 1988. "Public Experiments and Displays of Virtuosity: The Core-Set Revisited." *Social Studies of Science* 18 (4): 725–748.

Crommie, M. F., C. P. Lutz and D. M. Eigler. 1993. "Confinement of Electrons to Quantum Corrals on a Metal Surface." *Science* 262 (5131): 218–220.

Curtis, Scott. 2007. "The Rhetoric of the (Moving) Nano Image." Paper presented at "Images of the Nanoscale: From Creation to Consumption." University of South Carolina Nanocenter, Columbia, SC. October 26.

Daston, Lorraine, and Peter Galison. 1992. "The Image of Objectivity." *Representations* 40: 81–128.

Daston, Lorraine, and Peter Galison. 2007. *Objectivity*. Brooklyn, NY: Zone Books.

De Ridder-Vignone, Kathryn, and Michael Lynch. 2012. "Images and Imaginations: An Exploration of Nanotechnology Images." *Leonardo* 45 (5): 447–454.

Edgerton, Samuel Y. 1976. *The Renaissance Rediscovery of Linear Perspective*. New York: Harper & Row.

Elkins, James. 1995. "Art History and Images That Are Not Art." *Art Bulletin* 77 (4): 553–571.

Feynman, Richard. 1960. "There's Plenty of Room at the Bottom: An Invitation to Open Up a New Field of Physics." *Engineering and Science* 23 (5): 22–36.

Francoeur, Eric. 1997. "The Forgotten Tool: The Design and Use of Molecular Models." *Social Studies of Science* 27 (1): 7–40.

Frow, Emma. 2012. "Drawing a Line: Setting Guidelines for Digital Image Processing in Scientific Journal Articles." *Social Studies of Science* 42 (3): 369–392.

Frow, Emma. 2014. "In Images We Trust: Representation and Objectivity in the Digital Age." In *Representation in Scientific Practice Revisited*, edited by Catelijne Coopmans, Janet Vertesi, Michael Lynch and Steve Woolgar, 249–267. Cambridge, MA: MIT Press.

Garfinkel, Harold. 1967. *Studies in Ethnomethodology*. Englewood Cliffs, NJ: Prentice Hall.

Gilbert, G. Nigel, and Michael Mulkay. 1984. *Opening Pandora's Box: A Sociological Analysis of Scientists' Discourse*. Cambridge: Cambridge University Press.

Gombrich, E. H. 1960. *Art and Illusion: A Study in the Psychology of Pictorial Representation*. Washington, DC: The National Gallery of Art.

Goodsell, D. S. 2009. "Fact and Fantasy in Nanotech Imagery." *Leonardo* 42 (1): 52–57.

Hennig, Jochen. 2005. "Changes in the Design of Scanning Tunneling Microscopic Images from 1980 to 1990." *Techné: Research in Philosophy and Technology* 8 (2): 36–55.

Husserl, Edmund. 1964. *The Phenomenology of Internal Time-Consciousness*. Bloomington: Indiana University Press.

Husserl, Edmund. 1970. *The Crisis of European Sciences and Transcendental Phenomenology: An Introduction to Phenomenological Philosophy*. Evanston, IL: Northwestern University Press.

Husserl, Edmund. 1973. *Experience and Judgment: Investigations in a Genealogy of Logic*. Rev. ed. Evanston, IL: Northwestern University Press.

Kaplan, Sarah, and Joanna Radin. 2011. "Bounding an Emerging Technology: Para-scientific Media and the Drexler-Smalley Debate about Nanotechnology." *Social Studies of Science* 41 (5): 457–485.

Lösch, A. 2008. "Anticipating the Futures of Nanotechnology: Visionary Images as Means of Communication." In *The Yearbook of Nanotechnology in Society*, vol. 1, *Presenting Futures*, edited by Eric Fischer, Cynthia Selin and James M. Wetmore, 123–142. New York: Springer.

Lynch, Michael. 1988. "The Externalized Retina: Selection and Mathematization in the Visual Documentation of Objects in the Life Sciences." *Human Studies* 11 (2–3): 201–234.

Lynch, Michael. 1991. "Laboratory Space and the Technological Complex: An Investigation of Topical Contextures." *Science in Context* 4 (1): 51–78.

McLuhan, Marshall. 1964. *Understanding Media*. New York: Mentor.

Milburn, Colin. 2002. "Nanotechnology in the Age of Posthuman Engineering: Science Fiction as Science." *Configurations* 10 (2): 261–295.

Milburn, Colin. 2008. *Nanovision: Engineering the Future*. Durham, NC: Duke University Press.

Mody, Cyrus C. M. 2011. *Instrumental Community: Probe Microscopy and the Path to Nanotechnology*. Cambridge, MA: MIT Press.

Mody, Cyrus C. M., and Michael Lynch. 2010. "Test Objects and Other Epistemic Things: A History of a Nanoscale Object." *British Journal for the History of Science* 43 (3): 423–458.

Myers, Natasha. 2008. "Conjuring Machinic Life." *Spontaneous Generations* 2 (1): 112–121.

Myers, Natasha. 2014. "Rendering Machinic Life." In *Representation in Scientific Practice Revisited*, edited by Catelijne Coopmans, Janet Vertesi, Michael Lynch and Steve Woolgar, 153–175. Cambridge, MA: MIT Press.

Nerlich, Brigitte. 2005. "From Nautilus to Nanobo(a)ts: The Visual Construction of Nanoscience." *AZoNano: Online Journal of Nanotechnology* (December), DOI: 10.2240/azojono0109. Accessed March 11, 2014. http://www.azonano.com/article.aspx?ArticleID=1466.

Nerlich, Brigitte. 2008. "Powered by Imagination: Nanobots at the Science Photo Library." *Science as Culture* 17 (3): 269–293.

"New Cornell". 2003. "A New Cornell 'Nanoguitar,' Played by a Laser, Offers Promise of Applications in Electronics and Sensing."*Cornell News*, November 17, p. 30. Accessed August 9, 2012. http://www.news.cornell.edu/releases/nov03/nemsguitar.ws.html.

Nordmann, Alfred. 2004. "Nanotechnology's Worldview: New Space for Old Cosmologies." *IEEE Technology and Society Magazine* 23 (4): 48–54.

Ottino, Julio M. 2003. "Is a Picture Worth 1,000 Words? Exciting New Illustration Technologies Should Be Used with Care." *Nature* 421 (6922): 474–476.

Pitt, Joseph C. 2005. "When Is an Image Not an Image?" *Techné: Research in Philosophy and Technology* 8 (3): 24–33.

Polanyi, Michael. 1958. *Personal Knowledge: Toward a Post-Critical Philosophy*. London: Routledge & Kegan Paul.

Robinson, Chris. 2004. "Images in Nanoscience/Technology." In *Discovering the Nanoscale*, edited by Davis Baird, Alfred Nordmann and Joachim Schummer, 165–169. Amsterdam: IOS Press.

Robinson, Chris. 2012. "The Role of Images and Art in Nanotechnology." *Leonardo* 45 (5): 455–460.

Ruivenkamp, Martin, and Arie Rip. 2010. "Visualizing the Invisible Nanoscale: Visualization Practices in Nanotechnology Community of Practice." *Science Studies* 23 (1): 3–36.

Ruivenkamp, Martin, and Arie Rip. 2014. "Nanomages as Hybrid Monsters." In *Representation in Scientific Practice Revisited*, edited by Catelijne Coopmans, Janet Vertesi, Michael Lynch and Steve Woolgar, 177–200. Cambridge, MA: MIT Press.

Schummer, Joachim. 2006. "Gestalt Switch in Molecular Image Perception: The Aesthetic Origin of Molecular Nanotechnology in Supramolecular Chemistry." *Foundations of Chemistry* 8 (1): 53–72.

Sengupta, Samudra, Michael E. Ibele and Ayusman Sen. 2012. "Fantastic Voyage: Designing Self-Powered Nanorobots." *Angewandte Chemie* 51 (34): 8434–8445.

Thomas, Elisabeth Louise. 2004. *Emanuel Levinas: Ethics, Justice, and the Human beyond Being*. New York: Routledge.

Toumey, Chris. 2008. "Reading Feynman into Nanotechnology." *Techné: Research in Philosophy and Technology* 12 (3): 133–168.

Toumey, Chris. 2009. "Truth and Beauty at the Nanoscale." *Leonardo* 42 (2): 151–155.

6 Objectivity and Representative Practices across Artistic and Scientific Visualization

Chiara Ambrosio

VISUALIZATION, REPRESENTATION AND THE QUESTION OF OBJECTIVITY

Lorraine Daston and Peter Galison's (2007) work on scientific atlases has opened new lines of inquiry into the relations between objectivity and visual practices in science. Implicit in their narrative, and in need of further investigation, are some suggestions on how historical reflections on visual practices in science incorporate its evolving relation with the visual arts. In this chapter I chart the story of how artists participated in the practices of observation that Daston and Galison (2007, 19ff) compellingly define as "collective empiricism." In doing this, I use their narrative as a point of departure to narrate a story that is deeply intertwined with that of scientific objectivity, and which has remained so far largely untold. My aim is to show that the history of scientific objectivity has constantly crossed paths with the history of artistic visualization, from which it has received some powerful challenges. Drawing on three case studies from the eighteenth, nineteenth and twenty-first centuries, I argue that, by challenging the current canons of correct and accurate forms of visualization and representation, artists played a crucial role in shaping the history of objectivity—mainly by vocalizing their objections to it.

My concern in this chapter is twofold, and it ultimately aims to reconcile historical and epistemological accounts of visual practices in science and in the visual arts. At a basic level, my aim is to add new interpretative layers to images that are too often taken for granted and classified strictly as "scientific" or "artistic." Looking at the history of how certain artistic and scientific representations came to be the way they are reveals that scientific visualization is imbued with aesthetic commitments, and that artistic visualization constantly capitalizes on—and responds to—scientific and technological innovation. But in charting this history, my concern is primarily an epistemological one. Complicating the story of what counts as an accurate representation offers a challenge to current accounts of representation in philosophy of science. These accounts far too often qualify it as a relation between representing and represented facts that needs to be spelled out analytically.[1] While making sense of what *constitutes* representation is a central epistemological problem that deserves philosophical investigation, my claim is that representation, construed as an epistemological relation, should be explained in a historical fashion. The question is perhaps still one of discovery versus justification: Current philosophical accounts still concentrate on a normative quest for the formal constituents of representative relations, without paying much attention to the processes that led to their construction. What I want to stress, instead, is that representation is a relation that is *discovered* through processes and practices that are historically grounded, and that in many cases incorporate conversations and controversies between artistic and scientific ways of seeing.

Historicizing the category of representation also has the advantage of reinforcing its vital connection with visualization—a connection that is rarely addressed in current philosophical discussions. The atlas images discussed by Daston and Galison are a case in point, as they offer a glimpse of how collective practices of observation and shared styles of visualization become essential to "discern and stabilise" the objects of scientific investigation that appear on the atlas pages (Daston 2008, 98). In the course of this chapter, I add a new dimension to Daston and Galison's narrative, by showing that scientists were not alone in this process, and that in many cases they did not have the final word on what counts as an accurate representation. The story that I present is one about the contingencies that underpin what we usually regard as ready-made images, and the controversies that arise when artists and scientists respond to each other's modes of visualization.[2]

Throughout this chapter, I advocate a complementary view of visualization and representation, which I consider first and foremost as historically grounded practices and as constitutive components of experimentation and inquiry in science and in the visual arts. I conclude with a discussion of what a historicized view of representation has to offer epistemology and the philosophy of science. My final suggestion is to shift the focus of philosophical inquiry from a concept of "representation" to a historically grounded and pragmatic view of "representative practices," which better accounts for the key boundary areas in which art and science have complemented each other, and will continue to do so in the age of computerization.

Visualizing and Representing in the Eighteenth Century: "Truth-to-Nature"

> And as skeletons differ from one another, not only as to the age, sex, stature and perfection of the bones, but likewise in the marks of strength, beauty and make of the whole; I made choice of one that might discover signs of both strength and agility; the whole of it elegant, and at the same time not too delicate; so as neither to shew a juvenile or feminine roundness and slenderness, nor on the contrary

an unpolished roughness or clumsiness; in short, all of the parts of it beautiful and pleasing to the eye. For as I wanted to shew an example of nature, I chused to take it from the best pattern in nature. (Albinus [1747] 1749, sig. b.r.)

It is with these remarks that Bernhard Siegfried Albinus, professor of anatomy and surgery at the University of Leiden, introduced the illustrations that accompanied his impressive *Tabulae Sceleti et Musculorum Corporis Humani*, published in 1747. What Albinus was after, as a scientist, was an accurate representation of the human body, one that could take him beyond nature's variety and individual imperfections, and condense his philosophical ideal of "homo perfectus" (Punt 1983; Hildebrand 2005, 557). Albinus's perfect man required the choice of "the best pattern in nature." His ideal skeletons could not be too short, too tall or excessively slender, nor should they display the overly muscular appearance that was characteristic of the Baroque bodies featured in Andeas Vesalius's *De Humani Corporis Fabrica* two centuries earlier.[3]

Albinus's quest for an idealized image, which might or might not have had a concrete counterpart in nature, is only one of the examples that Daston and Galison discuss as part of what they define as "truth-to-nature" (2007, 70ff). This was the representative standard that eighteenth-century savants pursued before the term "objectivity" became the hallmark of accurate representations: "[Eighteenth-century] images were made to serve the ideal of truth—and often beauty along with truth—not that of objectivity, which did not yet exist" (Daston and Galison 2007, 104). Truth-to-nature imposed thorough familiarity with nature's variations, which served the purpose of "perfecting" individual occurrences as encountered in nature. Eighteenth-century sages had the scientific duty to correct nature for the sake of truth: Their illustrations show that for them representing was continuous with the act of discernment required to visualize phenomena not in their singularity but in their most general, ideal manifestations.

Very few scientists could achieve the ideal of truth-to-nature by themselves. For this purpose, eighteenth-century scientists needed to collaborate with artists—a collaboration that ultimately aimed at a fusion of the head of the scientist with the hand of the artist (Daston and Galison 2007, 88). But the story of artists and scientists working side by side, which Daston and Galison aptly describe as "four-eyed sight" (ibid.), is also one of conflict and controversies, of scientists enforcing ideas of perfection and truth and artists reacting in more or less overt, and often quite inventive, ways. While Daston and Galison pay some attention to this collaborative dimension, I suggest that much more can be learned from what remains hidden in the background of images that are usually perceived as sanctioning the victory of scientific accuracy (however construed) over artistic interpretation. It is here that the case of Albinus's collaborator, the Dutch artist and engraver Jan Wandelaar, offers an insightful example of the ways in which artists engaged with, and challenged, scientific modes of observation and visualization—with rather spectacular results.

Wandelaar's name is often mentioned by specialists as the hand behind Albinus's *Tabulae*.[4] Strangely enough, however, historians have paid scant attention to the accomplishments of Albinus's collaborator. Details of Wandelaar's artistic activity remain at present scattered throughout the literature, which mostly focuses on Albinus's innovations in the field of anatomy. Piecing together this fragmented information reveals that Wandelaar cultivated a range of scientific and intellectual interests that converged into his illustrations. In particular, Wandelaar's artistic practice was strongly driven by his knowledge of natural history, which he cultivated in parallel with his work as an illustrator and engraver of anatomical atlases. It was this knowledge that ultimately allowed him to give Albinus's *Tabulae* their distinctive character.

Following the death of his son, Wandelaar moved into Albinus's house and lived there for about twenty years. The two had met in 1723, when they were both working on a re-edition of Vesalius's *De Humanis Corporis Fabrica* (Huisman 1992, 2; Hildebrand 2005, 559). By the start of his work on the plates of Albinus's *Tabulae*, Wandelaar was already an accomplished illustrator and engraver. He had been collaborating with scientists at least since the 1720s, having designed and executed—among other things—the frontispiece of Linneus's *Hortus Cliffordianus*, published in 1737 (Daston and Galison 2007, 57). Under Albinus's strict guidance, Wandelaar perfected his scientific training, and his efforts to accommodate the scientist's requirements of accuracy and precision paid back. Albinus himself, in his memoirs, praised the artist's skills, his inquisitive mind and his willingness to learn human anatomy: "I have always wondered at his spirit, his patience and his resolution; he is moreover ardent and never without a certain impetuous eagerness of effort" (Huisman 1992, 2).[5]

With Wandelaar, Albinus had found a convenient way to avoid one of the major sources of inaccuracy that had affected anatomical representations until his time: the inevitable mismatch between anatomical visualization and its concrete representation in the final engraved image. The passage from anatomical preparations to drawings, and from these to copperplates or woodblocks, required the intervention of an engraver to transfer the artist's drawings on plates. But while the artist's eye and hand had been trained directly by the anatomist, the engraver entered this process only in its final stages—and this was the cause of most mistakes. Wandelaar could offer both services—drawing and engraving—thus securing the necessary continuity in the process of producing anatomical representations under the close supervision of Albinus himself.

Despite this, Albinus still demanded full control over his collaborator's work, imposing observational skills and methods to ensure that Wandelaar would learn the most accurate way of visualizing and perfecting anatomical features. The strictest control of the scientist over the artist was indispensable, in order to avoid the lack of accuracy that Albinus thought affected earlier anatomy atlases (to which the *Tabulae* were supposed to offer a direct corrective), such as Vesalius's *Fabrica* or Govard Bidloo's 1685 *Anatomia Humani Corporis* (Huisman 1992, 2).

A major concern for Albinus was the effect of distortion deriving from drawing in perspective. The problem was due to foreshortening, and to the fact that the artist could observe only part of the subject at a right angle. Thus, whatever happened to be at the center of the artist's field of vision, and viewed frontally, was depicted correctly. But the parts that were further removed from the center were observed—and consequently represented—at progressively sharper angles, with inevitable distortions and inaccuracies.

In an attempt to preserve both anatomical detail and correct representations, Albinus devised a two-step method that constituted a genuine innovation in the practice of anatomical illustration. This involved a system of double grids, or "diopters," which would allow the artist to maintain the proportions of the "homo perfectus," along with preserving the most accurate degree of detail at the right angle of observation (Huisman 1992, 3).[6]

The first step consisted in drawing the body or skeleton from a distance. The subject to be depicted was placed at 40 Rhenish feet (approximately 12.5 meters) from the artist. A first, large diopter, composed of a grid of squares each measuring 7.3 x 7.3 centimeters, was placed in front of the body. The artist drew on a sheet divided in the same manner as the diopter, at this stage focusing only on a general layout of the body, without introducing any detail (Huisman 1992, 4).

The second step consisted in filling in the initial life-size outline with anatomical details depicted with the utmost accuracy and precision, at the same time avoiding perspectival distortions. For this purpose, Albinus devised a smaller diopter, with a grid whose squares were one tenth of the size of the large diopter. Both diopters were placed in front of the body, with the smaller grid at a distance of 4 Rhenish feet from it. The smaller diopter had to correspond precisely with the squares forming the grid of the larger one, so that each 7.3 x 7.3 centimeters square on the larger grid would be divided in a hundred squares ten times as small. This system allowed Wandelaar to reproduce the details as if they were observed at 40 Rhenish feet without distortions (Punt 1983, 21–32).[7]

Historians have paid considerable attention to Albinus's method—and rightly so, as his experiments in measurement and accurate representation were the first of their kind in the eighteenth century. There is, however, an aspect of the *Tabulae* that has remained relatively neglected in the literature, and that constitutes their most striking feature: the backgrounds of the engravings. Albinus's idealized bodies are placed in floral, idyllic landscapes, and surrounded by neoclassical architectural elements. The images and symbols of vitality presented in the backgrounds were in part related to Albinus's ideas about the unity of nature (Hildebrand 2005, 561); however, they also constituted Wandelaar's hard-gained space for artistic expression.[8] It was in fact Wandelaar who had convinced Albinus about the importance of the backgrounds of the plates, which he justified as an effort "to preserve the proper light of the picture, for if the space around the figure and between its parts were white, the light would suffer" (Elkins 1986, 94).[9] Alas, Wandelaar's effort to preserve "the proper light of the picture" resulted in the two illustrations that were to become the symbol of Albinus's atlas: tables IV and VIII (Figures 6.1 and 6.2), in which the ideal skeleton is depicted with an equally well-proportioned rhinoceros in the background.

The animal in question is Clara, a female Indian rhinoceros that arrived in Leiden in 1741 and traveled through Europe between 1746 and 1756 (Clarke 1974, 116; Ridley, 2004; Rookmaaker 2005, 239). Clara is also known as the "Leiden rhinoceros," a name she was given after her Dutch owner, Captain Douvemont van der Meer of Leiden (Clarke 1974, 116). Historians usually explain the presence of Clara in the background of Albinus's illustrations as an exotic rarity that added an element of sophistication to the *Tabulae* (Hildebrand 2005, 561; Daston and Galison 2007, 72). This is partly true (Albinus himself substantiated this explanation in the text accompanying the plates), but there is more to the story.

What historians often neglect is that, by the publication of Albinus's *Tabulae* in 1747, Wandelaar had been drawing images of the rhinoceros for at least twenty years. In 1727 he was commissioned to illustrate the Dutch translation of the complete description of the geography, ethnography and natural history of Cape of Good Hope written by the German naturalist Peter Kolb. The work, originally published in German in 1719, devoted almost two folio pages to a description of a two-horned African black rhinoceros. Interestingly, the plates of the German edition conflicted with Kolb's description, in that they portrayed the rhinoceros according to a tradition dating back to Dührer's 1515 iconic representation of the animal. Following Dührer, most of these traditional illustrations, mainly based on second-hand accounts of the rhinoceros's features, presented the animal as one-horned (which is characteristic of the Indian species), covered by a thick armor and with smaller spurious horns on the shoulders (Rookmaaker 2005, 365ff).

Wandelaar's commission required him only to copy the plates from the 1719 edition of Kolb's book, but in fact he produced two different representations, both of which were eventually included in the final publication: a traditional, Dührer-like depiction of the rhinoceros, which matched the illustration included in the German edition, and a second image, which corrected and rectified it by following meticulously Kolb's description (Rookmaaker 1976, 88).[10] In this second illustration, Wandelaar's black rhinoceros is represented, probably for the first time, with smooth skin and two horns. With this bold move, Wandelaar became one of the first artists who broke with the established tradition that had dominated the iconography of the rhinoceros—and even affected its classification—for over two hundred years.[11]

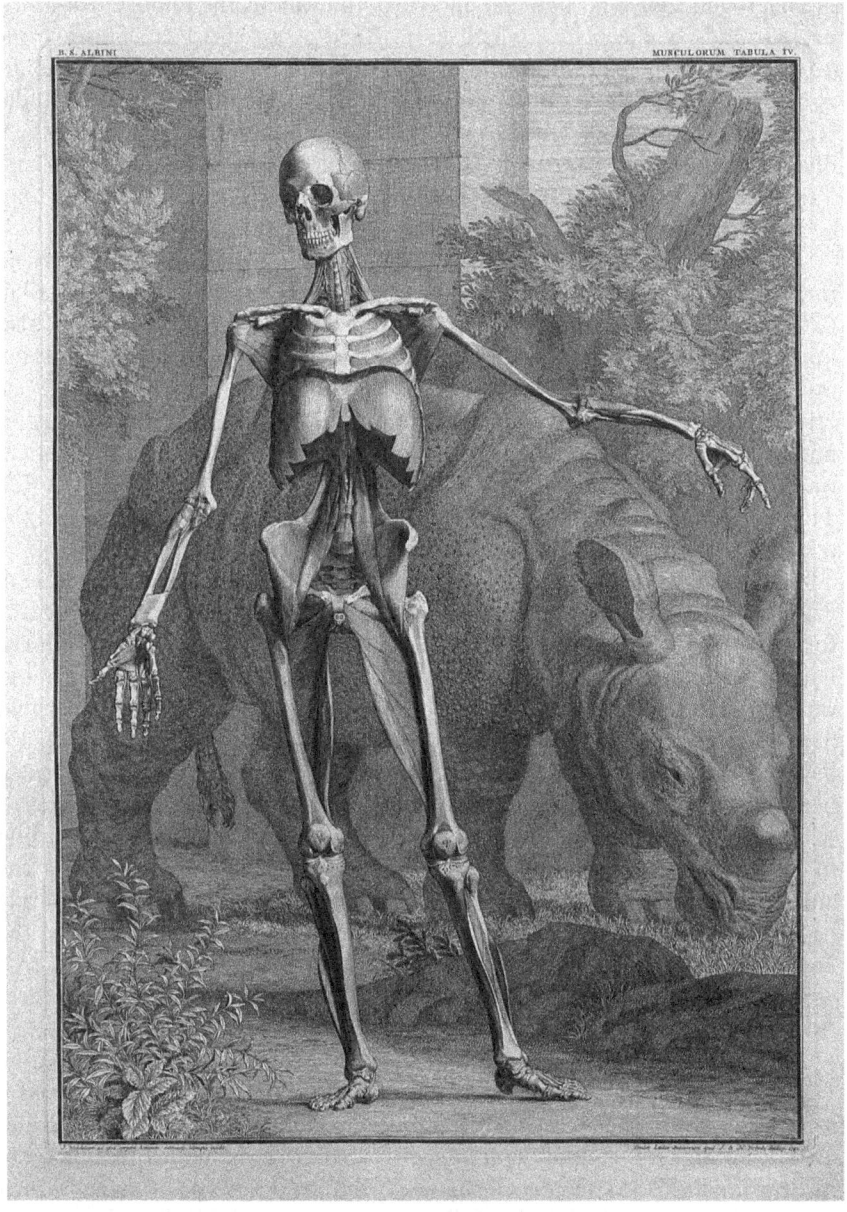

Figure 6.1 Bernard Siegfried Albinus, *Tabulae Sceleti et Musculorum Corporis Humani*, Plate IV. Wellcome Library, London.

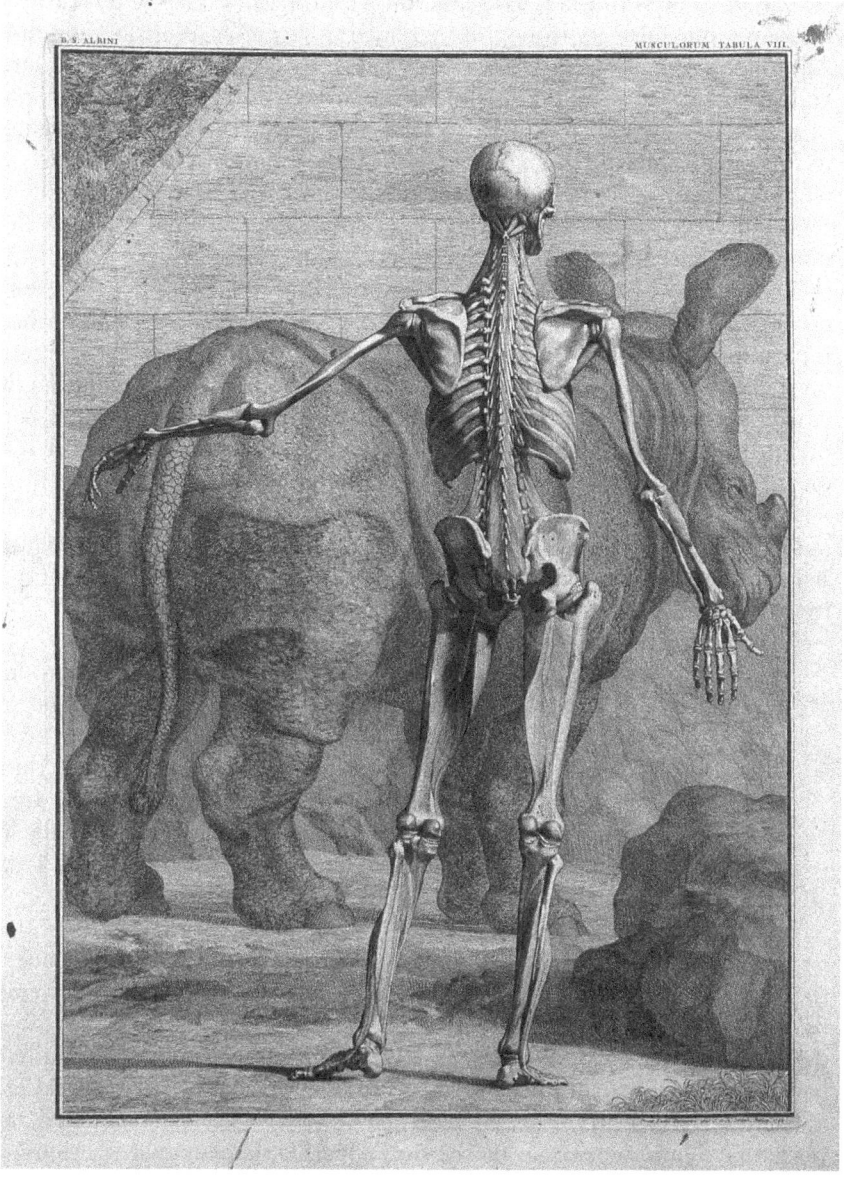

Figure 6.2 Bernard Siegfried Albinus, *Tabulae Sceleti et Musculorum Corporis Humani*, Plate VIII. Wellcome Library, London.

The story of Wandelaar's fascination with the rhinoceros and his role in the development of its iconography renders the presence of the animal in Albinus's *Tabulae* much less surprising. Far from being merely a fanciful and sophisticated addition to Albinus's anatomical works, the background of plates IV and VIII tells a story that runs parallel to the one presented in the foreground of the engravings—a story about the insightful ways in which artists occasionally interfered with the criteria for accurate representation imposed by scientists to pursue their own agendas, in ways that eventually became even scientifically acceptable.

The inclusion of the rhinoceros in Albinus's plates must have involved a great deal of negotiation. The fact that Clara belonged to the one-horned, Indian species of rhinoceros helped Wandelaar's cause, as her features were far more compatible with traditional representations of the animal than the two-horned African species he drew in 1727 (here one cannot avoid wondering whether Albinus would have dismissed a two-horned rhinoceros as one of nature's many exceptions!).

Hints of the reasons that Wandelaar might have used to persuade Albinus can be inferred from the commentary to the tables, where Albinus himself explicitly justifies the presence of the rhinoceros in the background of the plates as follows:

> We conclude this table, and the eight, by exhibiting in the back ground the figure of a female Rhinoceros that was shewed to us in the beginning of the year 1742, being two years and a half old, as the keepers reported. We thought the rarity of the beast would render these figures of it more agreeable than any other ornament resulting from mere fancy. The figures are just, and of a magnitude proportionable to the human figure contained in these two tables. (Albinus [1747] 1749, sig.g.1.v.)

Here Albinus's concern is maintaining the proportions: The presence of the rhinoceros in the plates is somehow justified by the fact that the depiction is "just" and "proportionable" to the human skeleton. At the same time, however, Albinus confesses between the lines that the plates partly betray his ideal of truth-to-nature. What one sees in the background is a particular rhinoceros, observed alive in 1742, when she was two and a half years old: Clara appears to be the only identifiable character in Albinus's *Tabulae*! With a brilliant operation of visual persuasion, Wandelaar managed to sidestep Albinus's quest for idealized types, and brought the particular right to the core of the two most representative illustrations of his anatomical atlas. Clara's cumbersome presence in the background of the plates vindicates the role of Wandelaar, and with him the crucial contributions of artists in shaping and challenging what counted as an accurate representation in the eighteenth century.

Daston and Galison's view that eighteenth-century "four-eyed sight" required the subordination of artists to scientists thus needs to be read with a few caveats. Some artists, like Wandelaar, approached accurate representation with all the dilemmas arising from the conflict between the needs of artistic experimentation and the impositions deriving from the canons of scientific accuracy. Hence, while the idealized bodies in the foreground of Albinus's illustrations are a sign of the ethos of discipline (well epitomized by the double grids and calculations) enforced by the scientist upon the artist, the rhinoceros in the background of plates IV and VIII shows that in some cases artists resisted the visual restrictions imposed by the pursuit of truth-to-nature. Ultimately, the story of Wandelaar's rhinoceros shows that eighteenth-century artists did not necessarily sacrifice their commitments for the greater cause of scientific accuracy: Instead, they approached scientific illustrations with their own visual priorities, leaving more or less visible traces of their presence in the pictures.

The tensions, conflicts and stratagems of persuasion that characterize the relationship between artists and scientists add a new dimension to Daston and Galison's narrative, and they are not restricted merely to the eighteenth century. In the mid-nineteenth century a new representational mode, that of mechanically reproduced images, saw artists and scientists openly arguing over the status of accurate representations. This controversy was openly construed under the heading of "objectivity," a term that entered the artist's vocabulary when photography made its first appearance in artistic practice.

Mechanical Objectivity: A View from Artistic Practice

Since 1839, when the first daguerreotype was presented at the Paris Académie des Sciences, the status of photography had been the subject of animated discussions. Scientists initially regarded it as the ultimate tool to obtain accuracy in observation and measurement. Its mechanical, reproducible and reliable nature was a reason to believe that it would function as "an artificial retina . . . at the disposal of the physicists," as Jean Baptiste Biot enthusiastically announced to the assembled members of the Académie des Sciences (quoted in Wilder 2009, 9). By the end of the nineteenth century, photography was used by scientists as an instrument of observation of phenomena that were considered otherwise unobservable, and it was employed as a form of measurement and as a means of obtaining experimental evidence (Didi-Hubermann 1986; Daston and Galison 2007; Wilder 2009).

Daston and Galison place the emergence of the modern concept of objectivity almost concomitantly with the birth of photography. Despite the fact that their concept of mechanical objectivity extends to a much broader range of recording instruments, photography constitutes a keystone in the process that led scientists to adopt an attitude of noninterventionism toward the objects of their inquiries:

> One type of mechanical image, the photograph, became the emblem for all aspects of noninterventionist objectivity . . . This was not because the photograph was more obviously faithful to nature than handmade images—many paintings bore a closer resemblance to their subject matter than early photographs, if only because they used color—but because the camera apparently eliminated human agency. (Daston and Galison 2007, 187)

Mechanical reproducibility contrasted sharply with the eighteenth-century ideal of truth-to-nature. Eighteenth-century representations required the willful intervention of the scientist—indeed, willful intervention was just what gave images their credibility and scientific status. Mechanical objectivity, on the other hand, required an attitude of asceticism (in its strongest moral connotations) toward the objects of scientific inquiry (Daston and Galison 2007, 120ff). Letting nature speak for itself became the nineteenth-century scientist's motto, human intervention being now replaced by a procedural use of image technologies that would ensure the elimination of the scientist's judgment from image-making, and even from the process of visualizing. This form of objectivity went hand in hand with the increased scientific reliance on recording instruments, which, like the camera, seemed to promise the possibility of removing human agency altogether.

Contrary to scientists, artists sought in photography a creative medium that could enhance willful intervention. The earliest stages of artistic photography saw artists treating the expressive possibilities offered by the camera as complementary and comparable to painting. Pictorialism, a movement that became dominant in the 1890s, explicitly aimed to differentiate artistic photography from scientific photography by treating the former as painting. Pictorialist photographers accomplished this by selecting carefully the content and the perspective from which photographs were taken, and intervened on the pictures by directly retouching them. This practice aimed to bring the artist's subjective intervention right at the core of technical photography—indeed, it aimed to stress the impossibility of removing agency from photography, no matter what scientists thought or how they used their photographic machinery.

Pictorialists' attacks against the supposedly objective nature of scientific photographs were open and frequent. Mechanical objectivity saw artists overtly reacting to science, in a manner that sharply contrasts with Wandelaar's somehow respectful intrusion in the background of Albinus's drawings. Open controversy often took sarcastic tones, as in a 1903 article aptly entitled "Ye Fakers," in which the pictorialist photographer Edward Steichen explicitly mocked the attitude of asceticism preached by the supporters of noninterventionist objectivity:

Some day there may be invented a machine that needs but to be wound up and sent roaming o'er the hill and dale, through fields and meadows, by babbling brooks and shady woods—in short, a machine that will discriminatingly select its subject and by means of a skillful arrangement of springs and screws, compose its motif, expose the plate, develop, print, and even mount and frame the result of its excursion, so that there will remain nothing for us to do but send it to the Royal Photographic Society's exhibition and gratefully to receive the Royal Medal. (Steichen 1903, 107)

Steichen's article first appeared in a journal with the explicit mission of advancing artistic photography as a form of art in its own right. The journal's name was *Camera Works*, and its founder, Alfred Stieglitz, occupies a central place in modernist responses to mechanical objectivity. Contrary to his contemporaries, Stieglitz did not reject objectivity altogether. By proposing an experimentalist aesthetics based on a view of the artist as a trained observer, Stieglitz challenged both the resistance to mechanical reproducibility pursued by pictorialist photographers and the noninterventionist attitude cultivated by scientists in the mid-nineteenth century.

Stieglitz is widely recognized as the pioneer of modernist photography and the key impresario of American avant-garde. A visionary intellectual and promoter of European modernist movements in America, he played a key role in defining the theoretical and practical foundations of photography as a form of art. What art historians often neglect, however, is that the foundations of his practice as a photographer lay in the climate of experimentalism that characterized German science in the 1880s.[12]

Born in Hoboken (New Jersey) from a German family, Stieglitz moved to Berlin in 1881 to enter the Charlottenburg Polytechnic and begin a degree in mechanical engineering. Stieglitz's biographers have often dismissed his student years as an initial obstacle to the development of his artistic career (Lowe 2002, 74). Yet the impact that his engineering and chemical training had on his subsequent artistic production tells a different story.

In the 1880s Berlin hosted a lively scientific community, which attracted the young Stieglitz since his early days at the Polytechnic. In parallel with his initial steps in the field of photography, he attended lectures by prominent figures such as the physicists Hermann von Helmholtz and Heinrich Hertz, the physiologist Emil DuBois-Reymond and the anthropologist and pathologist Rudolf Virchow (Kiefer 1991, 61ff; Lowe 2002, 73). But one of the figures who influenced Stieglitz in the most dramatic way, eventually compelling him to switch from engineering to chemistry, was the chemist August Hofmann.

Hofmann is especially known for his work on coal tar and his contribution to the development of aniline dyes, which laid the foundations of the German dye industry. A student of Justus Liebig at the University of Giessen, he had been, under his teacher's guidance, a pioneer in the transition from analytic to synthetic organic chemistry. Hofmann adopted and extended Liebig's successful methodology, whose distinctive trait was the integration of teaching and research in the practical setting of the chemical laboratory.[13] Since his early years under Liebig's guidance, Hofmann had structured his laboratory as a research community, in which chemical knowledge was conveyed through practice. Most of the daily learning happened by observing and doing, whereas lectures provided a theoretical background for students who lacked prior chemical training. Indeed, one of the most important methodological points that Hofmann adopted from Liebig was a systematic philosophy of chemical practice, whereby "experimental skill as well as theoretical convictions guided the analyst along a highly uncertain path from experiment to formula" (Jackson 2009, 16).

The concept that practice, far from being subordinate to theory, was constitutive of it became especially important to Stieglitz. The scientific aesthetics underpinning his practice as a photographer revolved around the idea that photography and science shared the same experimental basis and that in both cases theoretical considerations emerged as generalizations from practical experience. When, in 1905, Stieglitz established the Little Gallery at 291–293 Fifth Avenue, in New York, he characterized it as his "experimental station," an expression that was adopted by his closest collaborators:

> It should be remembered that the Little Gallery is nothing more than a laboratory, an experimental station, and must not be looked upon as an Art Gallery in the ordinary sense of the term. (De Zayas 1910, 47)

Stieglitz's breakthrough as the pioneer of modernist photography and as the impresario of avant-garde art in America consisted in adopting (and adapting) a Liebig-inspired model of a laboratory conceived as a social space with its community, collective observational practices, shared representational conventions and tacit ways of conveying knowledge through action. Moreover, just as a scientific research community, Stieglitz and his laboratory group disseminated their findings through the journal *Camera Work* (published between 1903 and 1917), which became one of the most important instruments for the promotion of avant-garde in the twentieth century (Eversole 2005).

Stieglitz's chemical training under Hofmann prevented him from unconditionally adopting the attitude of extreme interventionism that characterized pictorialist approaches to photography. While pictorialism still maintained a prominent place in *Camera Work* throughout the years of its publication, Stieglitz departed from it to embrace a more complex aesthetic position, which he identified as "straight photography." This new mode of visualization hinged on the role of trained observation, which Stieglitz considered as the main route to achieve objectivity through experimental inquiry. Stieglitz's concept of the "seer" behind the camera explicitly appeals to a scientific view of the trained eye, whose active judgment selects and interprets relevant aspects of a complex reality, and transposes them in a "true" photograph:

> It is high time that the stupidity and sham in pictorial photography be struck a solar plexus blow . . . Claims of art won't do. Let the photographer make a perfect photograph. And if he happens to be a lover of perfection and a seer, the resulting photograph will be straight and beautiful—a true photograph. (Stieglitz, in Adato 2001)

In this respect, I argue that Stieglitz's approach to photography anticipates the transition from the asceticism of mechanical objectivity to the community-informed ethos of inquiry that Daston and Galison (2007, 309ff) characterize as "trained judgment." Distinctive of twentieth-century image-making, trained judgment was a reaction to the constraints imposed by mechanical reproducibility. This new representational mode incorporated scientists' progressive awareness that judgment-inflected vision, rather than the "blind sight" of mechanical objectivity, was the goal of scientific visualization. Such an ethos of inquiry, which built interpretation in the process of image-making without depriving photographs of their "straight" character, was just what Stieglitz had cultivated within the experimental setting of his galleries.

One way of assessing this claim is through a work that constitutes a turning point in Stieglitz's aesthetics: his 1907 photograph *The Steerage* (Figure 6.3). The image condenses the artistic outcomes of his evolving views on scientific experimentalism and marks the culminating point of his synthesis of art and science.

The story of *The Steerage* is well known to art historians.[14] Stieglitz was traveling to Europe on board the liner *SS Wilhelm II*. Despite having a place on the first-class deck, he wandered with his camera in the vicinities of the steerage, with the purpose of taking pictures. In his memoirs, he recalls the taking of *The Steerage* as follows:

> A round straw hat, the funnel leaning left, the stairway leaning right, the white draw-bridge with its railings made of circular chains—white suspenders crossing on the back of a man in the steerage below, round shapes of iron machinery, a mast cutting into the sky, making a triangular shape. I stood spellbound for a while, looking and looking and looking. Could I photograph what I felt, looking, looking and still looking? I saw shapes related to each other. I saw a picture of shapes and underlying that the feeling I had about life. (Stieglitz, quoted in Norman 1942–1943, 128)

The Steerage is usually regarded by art historians as a photograph whose implications are mainly political and social. Without refuting this

Figure 6.3   Alfred Stieglitz, *The Steerage*, 1907 (gelatin silver print). The Israel Museum, Jerusalem, Israel. Gift of Augusta and Arnold Newman, New York, to American Friends of the Israel Museum. The Bridgeman Art Library. © Georgia O'Keeffe Museum / DACS, 2014.

interpretation, I would like to suggest that there is more to Stieglitz's photograph. My claim is that the visual effectiveness of *The Steerage* lies primarily in its conceptual nature—what Stieglitz compellingly defines as seeing "shapes related to each other." Conceptual relations are what Stieglitz was after in light of his scientific training. By concentrating on the inner relations between forms, in *The Steerage* Stieglitz obtained a photographic representation that verged on the conceptual. Far from exhibiting a faithful, point-to-point correspondence to a concrete event, *The Steerage* condenses Stieglitz's awareness that photography, as any form of representation, entails a process of abstraction and generalization from visual experience. His artistic quest for structure and form, for visualizing general properties beyond what was mechanically reproducible on the photographic plate, found its ultimate realization in *The Steerage*, and was modeled on his chemical laboratory practice.

Under Hofmann's guidance, Stieglitz had come to appreciate that chemical knowledge proceeds from experiment to general formulae. As the results of practical experimentation, chemical formulae are abstractions of the objects they stand for. At the same time, however, they are richer and more informative than their objects, for they capture structural properties of the experimental processes from which they arise. Moreover, by practicing chemistry in Hofmann's laboratory, Stieglitz had become familiar with the view that practice and process are constitutive components of theoretical knowledge. These basic principles allowed him to approach photography as a scientific problem to be solved experimentally, and devise a novel approach to photography that ultimately dispensed with exact correspondence as a criterion for representation.

With *The Steerage*, Stieglitz found a satisfactory balance between his photographic practice and his experimental philosophy. By challenging both naïve photographic realism and the simplistic appeal to subjectivity pursued by pictorialist photographers, he formulated a novel relation at the basis of his photographic representations, one that was built on the trained observer's active judgment in visualizing and selecting salient properties from the objects of photographic inquiry.

Stieglitz was well aware that photography, as every act of observation, is theory-dependent. And the theory that informs photography is in turn shaped by the needs and goals of the photographer, along with his tools, chemical equipment and laboratory practice. His scientific training provided him with a renewed awareness of this aspect of photography and of the experimental process that guides the photographer from "looking, looking and still looking" to the final image—a "true" photograph. This characterization of Stieglitz's approach to photography suggests that his formulation of objectivity was not at all concerned with mechanical reproducibility. Far from preaching restraint and asceticism, Stieglitz recognized that objectivity required a trained "seer," and that the informed activity of trained observers constituted the crucial connection between artistic visualization and mechanically reproduced images.

Historians and philosophers of science far too often tend to confine the changes that affected scientists' views about objectivity—especially with respect to mechanical reproducibility—only within science. But then there was art. There were pictorialist photographers who reacted to photographic objectivity with the compelling means of artistic experimentation and singled themselves out because of their obstinacy with the subjective aspects of artistic practice against the attitude of noninterventionism preached by scientists. And there were scientifically minded artists, such as Stieglitz, who sought for a compromise between resistance and restraint. In this process, scientific visualization was a parameter that artists had constantly in mind: Whether they adopted it or reacted to it, science informed their visual inquiries into the ways in which representations capture perspicuously some aspect of the world.

ARTISTIC VISUALIZATION AS "CRITIQUE"

Daston and Galison's narrative stops at a point that marks a crucial junction in the history of visualization: the contemporary shift from "representation" to "presentation" (Daston and Galison 2007, 382ff). In the age of computerization, visualization challenges the boundaries between the artifactual and the natural: The new scientific images fulfill the purpose of manipulating the real—and they do so in an aesthetically pleasing way. This new representational mode opens a new chapter in the story of the relations between artistic and scientific visual practices: Far from being disciplined and restrained, artists are now invited to take a central place in the scientific enterprise, as privileged instruments of scientific visualization. This suggests that the time of controversy, of artists having to hide themselves in the background of anatomical engravings or having to engage in arguments against the reliability of mechanically reproduced images, is finally over.

Such an optimistic view needs to be taken with a grain of salt. Several contemporary collaborations between artists and scientists still tend to relegate the role of artists as mere accessories in the toolkit of scientific visualization—a passive role that requires their vision to be disciplined in some way. A parallel (and equally perplexing) assumption is that the artist's work is a mere means to add a visually pleasant dimension to scientific visualization, which would magically render science more communicable.

What I would like to present in this section is an alternative view, which stresses once more the critical role of the artist in challenging what scientists take as granted, objective and unquestionable. My view is rooted in history: The case studies I have presented so far show that looking at the controversies that divided artists and scientists over what counts as accurate representation contributes an entirely new layer to the story of objectivity. At the same time, a normative lesson can be drawn from this history: Artists *should* continue to pursue their critical mission—in fact they should take it very seriously, because scientific practice benefits from it. As most normative statements, this is not unproblematic. It can be argued, for example, that not all artists indiscriminately would agree on having this mission. Additionally, other actors apart from scientists might benefit from the critical role of art (the artists themselves, or the public, when these collaborations are aimed at "engagement"). My normative claim here focuses on what scientific objectivity has to gain from involving artists, perhaps more explicitly than in the past, in the practice of scientific visualization. In this respect, I propose that the critical mission of art shares some of the features that characterize the role of philosophy of science: that of questioning and challenging assumptions and modes of working that would otherwise be taken for granted by scientific practitioners.[15]

My final case study exemplifies this point. I draw on a 2009 work by the artist Martin John Callanan, entitled *A Planetary Order (Global Terrestrial Cloud)* (Figure 6.4).

Callanan's artwork is a physical visualization of real-time "raw" scientific data. It captures at a glance one second of readings, taken on February 2, 2009, at exactly 0600 UTC, from all the six cloud-monitoring satellites overseen by NASA and the European Space Agency. The readings were

Figure 6.4 Martin John Callanan, 2009. *A Planetary Order (Global Terrestrial Cloud)*. Courtesy of the artist.

transformed into a 3-D computer model, which was then 3-D printed at the Digital Manufacturing Centre at the UCL Bartlett Faculty of the Built Environment (Hamblyn and Callanan 2009, 67).

The layer of clouds that covers the globe's surface, and that only suggests the presence of continents underneath, creates a powerful perceptual shift: Patterns of clouds that seem transient and mutable when viewed from earth form a coherent "planetary order" when seen from space. *A Planetary Order* gives a visible form to information that would have otherwise remained in the form of silent quantitative readings, whose self-evidence is generally accepted with no reservations by scientists. As a representation, the globe is not so much—or at least not exclusively—about the exact position of patterns of cloud over the Earth's surface. The piece seems instead to raise more fundamental questions about what count as supposedly "raw" data, and the various ways in which such data can be visualized.

Callanan's work was part of a broader project, carried out in 2009 in collaboration with the writer Richard Hamblyn. The project involved them respectively as an artist and a writer in residence working in an interdisciplinary team of researchers at the UCL Environment Institute. Their collaboration aimed to use artistic visualization as a critique of the scientific rhetoric of "data." Their results converged in a book, aptly entitled *Data Soliloquies* (2009), which features Callanan's *A Planetary Order* as one of its most iconic illustrations, and which draws on the visual narratives of climate change to underline more broadly "the extraordinary cultural fluidity of scientific data" (Hamblyn and Callanan 2009, 13). The book has far reaching implications about the ways in which scientific data are communicated to the public, their often spectacular modes of representation and the mechanisms of persuasion that are implicitly built in the display of large quantities of information. But a striking—albeit perhaps less obvious—aspect of the book is that it addresses, partly through Callanan's visual work, some important epistemological questions concerning the relation between data and their representation.

From the outset, Hamblyn and Callanan state that their aim is to "interrogate" data and bring to the fore the assumptions that far too often remain hidden behind their supposed self-evidence:

> Our title, "Data Soliloquies" . . . reflects the ways in which scientific graphs and images often have powerful stories to tell, carrying much in the way of overt and implied narrative content; but also that these stories or narratives are rarely interrupted or interrogated. They are information monologues—soliloquies—displayed more for their visual and rhetorical eloquence than for their complex (and usually hard-won) analytical content. (Hamblyn and Callanan 2009, 14)

Far from being "found objects," data are collected and constructed according to specific acts of judgment, which form intricate narratives and stories behind their visual immediacy and power. Hamblyn and Callanan's message is that the quantity of data available does not justify their self-evidence: Data visualization involves first and foremost an act of discernment and choice, in which patterns are cautiously carved out of the inevitable statistical uncertainty that surrounds them.

Along with the shift from representation to presentation, which Daston and Galison identify as the latest frontier of correct depiction, the rhetoric of scientific objectivity has progressively turned to data and the statistical correlations between them as a guarantee of scientific reliability.[16] This new appeal to objectivity hinges on quantity—a "data deluge"—without taking into account that its visual manifestations are themselves informed by judgment, discernment and choice. The task of the artist—well exemplified by Hamblyn and Callanan's work—consists in giving a visible form to such an act of discernment, and this places artistic visualization once again in the uncomfortable position of serving as a challenge to the parameters that define scientific objectivity.

Artistic visualization as "critique" consists in bringing judgment and uncertainty to the fore, in questioning the self-evidence of scientific data and interrogating scientific visualization, rather than letting it flow as an uninterrupted and undisturbed soliloquy. Untied from the discipline of truth-to-nature and the restraints of mechanical objectivity, artists can now vocalize their objections to objectivity in ways that can be immediately heard by scientists, and that can explicitly feed into their practice.

The kind of attitude toward objectivity—or at least toward the supposed objectivity of scientific data—that these new controversies will open is still difficult to anticipate. But the fact that artistic practice is now looking at data—a foundational and still little questioned assumption in scientific practice—is a promise that artists' challenges to scientific modes of visualization will not spare the fundamentals.

HISTORICIZING REPRESENTATIONS:
A PLEA FOR "REPRESENTATIVE PRACTICES"

> All epistemology begins in fear—fear that the world is too labyrinthine to be threaded by reason; fear that the senses are too feeble and the intellect too frail; fear that memory fades, even between adjacent steps of a mathematical demonstration; fear that authority and convention blind; fear that God may keep secrets or demons deceive. Objectivity is a chapter in the history of intellectual fear, of errors anxiously anticipated and precautions taken. (Daston and Galison 2007, 372)

Representations have not been spared from the history of intellectual fear, which features objectivity as one of its central chapters. Reduced to formalisms and logical relations, marginalized by philosophers as dispensable accessories of theories, representations have suffered the same fate as objectivity—perhaps because they constitute its most tangible materialization.

I began this chapter stating that my concerns about representation and visualization were mainly epistemological, and it is to those concerns that I want to return in my conclusions—even at the risk of adding a new chapter to the history of intellectual fear. In charting the story of how artists participated in the conversations and controversies surrounding accurate representations, I stressed that the history of scientific objectivity has constantly crossed paths with the history of artistic visualization. Whether glaringly displayed on the pages of books, atlases and journals, or hidden in the background of eighteenth-century engravings, artists' reactions to objectivity shaped its history, and with it the history of representation. The views I presented in the course of this chapter pose some challenges to current accounts of representation in philosophy of science. I want to offer two suggestions, which aim at providing a view of representations that reconciles epistemological and historical accounts, at the same time reconnecting them to the visual history of objectivity.

My first suggestion is that representations, even when conceptualized as epistemological relations, are inherently historical. This might seem a trivial point, but it is one rarely addressed by philosophers of science. The only way we have to make sense of the relation of representation—be it one of resemblance, denotation, idealization, abstraction or any of the various shades in between—is by looking at the processes that led to its construction.

The three case studies that I presented in this chapter are a case in point. The interplay between background and foreground in Albinus's tables shows that his collaborator, Jan Wandelaar, was torn between idealization—the criterion imposed by truth-to-nature—and particular representations. Wandelaar's epistemological dilemmas converged in tables IV and VIII, eventually shaping the distinctive representative character of the illustrations in Albinus's work. Stieglitz's conceptual approach to photography brought him beyond the realm of figurative depiction. Straight photography allowed him to move beyond a quest for a faithful correspondence between the image and its objects, and place the judgment of a trained "seer" right at the center of what counts as a true picture. Lastly, Callanan's *A Planetary Order* constructs representation as a translation of numerical data deriving from cloud readings into their 3-D, visual and tangible counterpart. Here figurative or conceptual representations are beside the point: What counts is the materialization of large datasets, their transposition into a material artifact that stands for the very artifactual nature of data. All three artists, despite belonging to very different historical periods, are concerned with the relation between their representations and the objects that such representations are meant to capture. Yet what links these three very different forms of artistic experimentation is not the presence of a single, permanently fixed, representative relation: If anything, all these works suggest that representation consists of a *plurality of relations*. What brings these three artists together is their common quest for an alternative to current modes of scientific visualization, of which their representations are a material incarnation. This quest is first and foremost an expression of how historical contingencies, arguments and controversies between artists and scientists ultimately feed into the relation of representation, and shape its final form.

Historicizing representations ultimately amounts to bringing to the fore not just the reasons but also the material and experimental practices involved in their construction. This leads me to a second and final suggestion: a plea to shift the focus of philosophical inquiry from "representation" to "representative practices," construed as the ways in which both artists and scientists devise useful and perspicuous ways of visualizing phenomena and experimenting upon them.[17] Without denying the theoretical components that guide visualization, the concept of representative practices captures at the same time the procedural and experimental aspects of representing, which far too often remain implicit in epistemological accounts. The case studies that I have presented in the course of this chapter show that artists' and scientists' quest for "accurate" or "correct" ways of representing invokes questions about what counts as a representation in the first place, and yet the ways in which they arrive at such representations reflect a journey of visual discovery, which parallels their concrete experimentation upon phenomena. Thus construed, the shift to representative practices offers a useful means of reconciliation for two dominant strands in the literature on representations in philosophy of science.[18] On the one hand, the concept of representation has been used by philosophers of science, especially within the analytic tradition, to investigate the relation between theory and the world. This approach suffers from the overemphasis on theories and the bias toward the physical sciences that characterized most philosophy of science at least until the early 1980s. Its immediate drawback is that it tends to dismiss experimental practices as merely subservient to theories. Being concerned mainly with problems of meaning, reference and the metaphysics of relations, the proponents of analytical accounts focus exclusively on what *constitute*s a representation, and ultimately view it as a relation between a source (the vehicle of the representation) and a target (the object of the representation), which needs to be spelled out in terms of necessary and sufficient conditions (Suárez 2010, 91ff).

The limits of such accounts have been partly recognized by philosophers of science. This has determined a recent shift in the literature on representation, which is in line with the contemporary tendency to integrate history and philosophy of science in a unified account. Rather than focusing on the nature of the relation of representation, recent accounts concentrate on *uses* and *means* as opposed to the *constituents* of representations. This approach draws on a more general turn toward practice in philosophy of science, inaugurated by Hacking's (1983) *Representing and Intervening*, which inverted the hierarchical relation between theory and experiment, placing the latter at the center of philosophical investigation. This was paralleled by a renewed interest in the role of representations—and more specifically the role of modeling practices—in scientific experimentation.[19]

Practice-based accounts focus on model building, rather than representation as a relation that needs to be spelled out analytically, as fundamental to scientific practice. Their starting assumption is that representations should be investigated in conjunction with the various uses of models and the particular contexts in which they are used. But despite its emphasis on the experimental contexts surrounding the use of models, there is a sense in which this new strand in philosophy of science had the effect of divorcing the theoretical implications of representations from their experimental uses and applications. The concern here is that the overemphasis on *use* in the literature on models tends to reduce them to ready-made entities, whose importance lies exclusively in their practical applications. In this, the proponents of practice-based accounts share the same difficulties as their analytical predecessors: Both seem to focus on the *justification* of the representative relation or of the ways in which we use representations, without explaining the process of discovery that is at stake in producing them. What I want to stress by appealing to a concept of "representative practices," instead, is that using representations is coextensive with the process of constructing them.

Once again, the three case studies that I have presented in this chapter are a case in point. The stories behind Wandelaar's rhinoceros, Stieglitz's *Steerage* and Callanan's *A Planetary Order* are cases of visualization construed as an experimental practice. They incorporate actions and physical processes—engraving, printing, laser-melting in 3-D—and serve practical purposes—observing, visualizing, experimenting with data, constructing persuasive arguments. But, at the same time, all these practical actions also incorporate and convey a fundamental, theoretical quest for what counts as a representation in the first place. Thus construed, they should be interpreted as multilayered arguments, rather than formalized relations or ready-made models fulfilling a single purpose.

Representative practices account for the key boundary areas in which art and science have complemented each other, and continue to do so in the age of computerization. They account for the process of experimental inquiry, both theoretical and practical, that has characterized the conversations and controversies between artists and scientists around the status and purposes of accurate representations, as well as around what representing amounts to in the first place. The history of objectivity tells us that these questions have shifted through time and, with them, so has the relation of representation. Truth-to-nature, mechanical reproducibility, expert judgment and the last frontier of representing large datasets: All the twists and turns that characterize the history of objectivity contribute to place a plurality of different relations at the basis of what counts as a representation. Such plurality is nourished and invigorated by the fact that artists, as much as scientists, do not divorce their theoretical questions about representations from their practical uses. It is in this sense that representative practices are part and parcel of the story of objectivity—and hopefully this account leaves only a note in the margins, rather than writing a new chapter, in the history of intellectual fear.

ACKNOWLEDGMENTS

I am grateful to Annamaria Carusi, Aud Sissel Hoel, Timothy Webmoor, Steve Woolgar and an anonymous referee for their perceptive comments on all the preliminary drafts of this chapter. I owe a great deal to Lorraine Daston and Peter Galison for their kind encouragement and for engaging in several conversations that proved invaluable in shaping my argument. Niall Le Mage provided sharp editorial criticism on the final draft of this chapter, and I greatly benefited from his comments. Paola Fumagalli at the Bridgeman Art Library offered invaluable advice on image copyright and licensing. The students in the courses HPSC 3022/HPSC3034: Science, Art, and Philosophy followed the development of this chapter since its earliest stages. Their enthusiasm, comments and questions contributed to give my story its present form.

NOTES

1. An overview of the current debate on the nature of scientific representations is in Suárez (2010).
2. Here I implicitly draw on the rich STS literature on scientific controversy, which includes Latour (1987), Collins and Pinch (1993) and more recently Castel and Sismondo (2008).
3. Even more importantly, Albinus's "homo perfectus" was a *man*, and as such his skeleton should definitely not display any roundness or frailty that was typically female. Women entered anatomical classification only toward the end of the eighteenth century, with Samuel Thomas von Soemmering claiming to have published the first illustration of a female skeleton. Schiebinger (1986) gives an account of the political and ideological implications of Soemmering's highly idealized female skeleton.
4. See, for example, Choulant (1962), Ferguson (1989), Elkins (1986) and Huisman (1992).
5. Huisman reports this quotation from the preface to Book 1 of Albinus's *Academicarum Annotationum*.
6. The double grids system had been devised for Albinus by Willem's Gravesande, whose *Essai de Perspective* had been published in 1711 (Elkins 1986, 94).
7. Historians diverge on how exactly Albinus used the second grid. Huisman (1992, 6ff) recalculates the size of the second diopter and adds that the life-size drawing thus obtained was further scaled to folio size by transferring the grid of 7.3 x 7.3 centimeters square to one with a further grid of 2.5x 2.5 centimeters.
8. Wandelaar eventually gained complete freedom as to what to include in the backgrounds. For instance, in the *Tabula Musculorum II*, a muscleman is

presented in front of a stairway with a sculpted lion in the background. I am grateful to Andrea Fredericksen, curator of the UCL Art Museum, for bringing this engraving to my attention. The table, from the 1749 English translation of Albinus's *Tabulae*, is part of the UCL Art Collections.

9. Also quoted in Ferguson (1989, 232). Both report Albinus's quote (originally in the *Academicarum Annotationum*) from earlier editions of Choulant (1962).

10. Wandelaar's depiction of the rhinoceros here crosses paths with the taxonomical concerns surrounding the classification of the animal and its representation in the eighteenth century. Only in 1758 Linneus provided the name *Diceros bicornis* for the two-horned African rhinoceros in his *Systema Naturae*, and even then the animal remained pretty much unknown in Europe. Most depictions continued to be based on specimens of one-horned Indian *Rhinoceros Unicornis*, which were best known in Europe. See Rookmaaker (2005) and Hanson (2010).

11. In the years following Kolb's publication, Wandelaar found more opportunities to cultivate his fascination for the rhinoceros. Two pages of sketches illustrating the details of a preserved specimen probably displayed at the botanical gardens in Leiden show that between 1734 and 1735 Wandelaar continued to work on his representations of the animal (Rookmaaker 1976, 88).

12. The only exception to these accounts is Kiefer (1991).

13. On Liebig's laboratory see Holmes (1989) and Jackson (2009). On Hofmann's adoption of Liebig's model see Bentley (1970, 1972) and Meinel (1992).

14. For a recent account of Stieglitz's 1907 photograph, see Francisco and McCauley (2012).

15. I am grateful to an anonymous referee for challenging this point in my discussion and allowing me to elaborate on it further (a clear case in which "critique" has benefits!).

16. In a well-known 2008 *Wired Magazine* article, for instance, Chris Anderson prophesized "the end of theories" as a result of the data deluge: "The new availability of huge amounts of data, along with the statistical tools to crunch these numbers, offers a whole new way of understanding the world. Correlation supersedes causation, and science can advance even without coherent models, unified theories, or really any mechanistic explanation at all" (Anderson 2008).

17. I have sketched a formulation of representative practices, whose roots are in the Pragmatist account of representations proposed by Charles S. Peirce, in Ambrosio (forthcoming).

18. In drawing this distinction I follow Suárez (2010).

19. The forerunners of this approach are Achinstein (1968), Black (1962) and Hesse (1963). Cartwright (1983) and Morrison and Morgan (1999) pioneered the shift toward practice-based approaches to models, whereas De Chadarevian and Hopwood (2004) have recently examined the historical context surrounding the production and use of models. More recent accounts include Suárez (1999, 2003), and useful overviews of this debate are in Frigg and Hunter (2010) and Suárez (2010). The turn to practice in philosophy of science has a sociological counterpart in the accounts of representation in practice presented in Lynch and Woolgar (1990).

REFERENCES

Achinstein, Peter. 1968. *Concepts of Science: A Philosophical Analysis*. Baltimore: John Hopkins University Press.

Adato, Perry Miller. 2001. *American Masters—A. Stieglitz: The Eloquent Eye*. New York: Winstar TV and Video.

Albinus, Bernhard Siegfried. (1747) 1749. *Tabulae Sceleti et Musculorum Corporis Humani*. Leiden: J. & H. Verbeck. Translated as *Tables of the Skeleton and Muscles of the Human Body*. London: John & Paul Knapton.

Ambrosio, Chiara. Forthcoming. *Iconic Representation, Creativity and Discovery in Art and Science*. Edited by Wenceslao Gonzalez. A Coruña: Netbiblo.

Anderson, Chris. 2008. "The End of Theory". *Wired Magazine*, July 23. http://www.wired.com/science/discoveries/magazine/16–07/pb_theory. Last accessed 4 November 2013.

Bentley, Jonathan. 1970. "The Chemical Department of the Royal School of Mines: Its Origins and Development under A. W. Hofmann." *Ambix* 17: 153–181.

Bentley, Jonathan. 1972. "Hofmann's Return to Germany from the Royal College of Chemistry." *Ambix* 19: 197–203.

Black, Max. 1962. *Models and Metaphors*. Ithaca: Cornell University Press.

Cartwright, Nancy. 1983. *How the Laws of Physics Lie*. Oxford: Oxford University Press.

Castel, Borris, and Sergio Sismondo. 2008. *The Art of Science*. Toronto: University of Toronto Press.

Choulant, Ludwig. 1962. *History and Bibliography of Anatomic Illustration*. Translated and annotated by Mortimer Frank. New York: Hafner.

Clarke, T. H. 1974. "The Iconography of the Rhinoceros. Part II: The Leyden Rhinoceros." *Connoisseur*, February, 113–122.

Collins, Harry, and Trevor Pinch. 1993. *The Golem: What Everyone Should Know about Science*. Cambridge: Cambridge University Press.

Daston, Lorraine. 2008. "On Scientific Observation." *Isis* 99 (1): 97–110.

Daston, Lorraine, and Peter Galison. 2007. *Objectivity*. New York: Zone Books.

De Chadarevian, Soraya, and Nick Hopwood, eds. 2004. *Models: The Third Dimension of Science*. Stanford, CA: Stanford University Press.

De Zayas, Marius. 1910. "Photo Secession Notes." *Camera Works* 30: 47.

Didi-Hubermann, Georges. 1986. "Photography—Scientific and Pseudo-Scientific." In *A History of Photography: Social and Cultural Perspectives*, edited by Jean-Claude Lamagny and André Rouillé, 71–76. Cambridge: Cambridge University Press.

Elkins, James. 1986. "Two Conceptions of the Human Form: Bernhard Siegfried Albinus and Andreas Vesalius." *Artibus et Historiae* 7 (14): 91–106.

Eversole, Ted. 2005. "Alfred Stieglitz's *Camera Work*, and the Early Cultivation of American Modernism." *Journal of American Studies of Turkey* 22: 5–18.

Ferguson, J. P. 1989. "The Skeleton and the Rhinoceros." *Proceedings of the Royal College of Physicians of Edinburgh* 19 (2): 231–232.

Francisco, Jason, and Elizabeth Anne McCauley. 2012. *The Steerage and Alfred Stieglitz*. Berkeley: University of California Press.

Frigg, Roman, and Matthew Hunter, eds. 2010. *Beyond Mimesis and Convention: Representation in Art and Science*. Dordrecht: Springer.

Hacking, Ian. 1983. *Representing and Intervening*. Cambridge: Cambridge University Press.

Hamblyn, Richard, and Martin John Callanan. 2009. *Data Soliloquies*. London: UCL Environment Institute.

Hanson, Craig Ashley. 2010. "Representing the Rhino: The Royal Society between Art and Science in the Eighteenth Century." *Journal of Eighteenth-Century Studies* 33 (4): 545–566.

Hesse, Mary. 1963. *Models and Analogies in Science*. London: Sheed and Ward.

Hildebrand, Reinhard. 2005. "Attic Perfection in Anatomy: Bernhard Siegfried Albinus (1697–1770) and Samuel Thomas Soemmering (1755–1830)." *Annals of Anatomy* 187: 555–573.

Holmes, Frederic Lawrence. 1989. "The Complementarity of Teaching and Research in Liebig's Laboratory." *Osiris* (2nd series) 5: 121–164.

Huisman, Tim. 1992. "Squares and Diopters: The Drawing System of a Famous Anatomical Atlas." *Tractrix* 4: 1–11.

Jackson, Catherine M. 2009. "Analysis and Synthesis in Nineteenth-Century Chemistry." PhD dissertation, University of London.

Kiefer, Geraldine. 1991. *Alfred Stieglitz: Scientist, Photographer and Avatar of Modernism, 1880–1913.* New York: Garland.

Latour, Bruno. 1987. *Science in Action: How to Follow Scientists and Engineers through Society.* Cambridge, MA: Harvard University Press.

Lowe, Sue Davidson. 2002. *Alfred Stieglitz: A Memoir/Biography.* 2nd ed. Boston: Museum of Fine Art.

Lynch, Mike, and Steve Woolgar. 1990. *Representation in Scientific Practice.* Cambridge, MA: MIT Press.

Meinel, Christopher. 1992. "August Wilhelm Hofmann—'Reigning Chemist-in-Chief.'" *Angewandte Chemie* 31 (10): 1265–1398.

Meinel, Christopher. 2004. "Molecules and Croquet Balls." In *Models: The Third Dimension of Science,* edited by Soraya de Chadarevian and Nick Hopwood, 242–275. Stanford, CA: Stanford University Press.

Morgan, Mary and Morrison, Margaret. 1999. Models as Mediators. Perspectives on Natural and Social Science. Cambridge: Cambridge University Press.

Norman, Dorothy, ed. 1942–1943. "Alfred Stieglitz; Four Happenings." *Twice a Year* 8–11: 105–137.

Norman, Dorothy. 1973. *Alfred Stieglitz: An American Seer.* New York: Random House.

Punt, Hendrik. 1983. *Bernard Siegfried Albinus (1697–1770), On "Human Nature." Anatomical and Physiological Ideas in Eighteenth-Century Leyden.* Amsterdam: B. M. Israël.

Ridley, Glynis. 2004. *Clara's Grand Tour: Travels with a Rhinoceros in Eigteenth-Century Europe.* New York: Atlantic Monthly.

Rookmaaker, L. C. 1976. "An Early Engraving of the Black Rhinoceros (*Diceros bicornis (L.)*) Made by Jan Wandelaar." *Journal of the Linnean Society* 8: 87–90.

Rookmaaker, L. C. 2005. "Review of the European Perception of the African Rhinoceros." *Journal of the Zoological Society of London* 265: 365–376.

Schiebinger, Londa. 1986. "Skeletons in the Closet: The First Illustration of the Female Skeleton in Eighteenth-Century Anatomy." *Representations* 14: 42–82.

Steichen, Edward. 1903. "Ye Fakers." *Camera Works* 1. Reprinted in *Alfred Stieglitz. Camera Work: The Complete Illustrations 1903–1917,* edited by S. Philippi and U. Kyeseyer, 107. Cologne: Taschen, 1997.

Suárez, M. 1999. "Theories, Models, and Representations." In *Model-Based Reasoning in Scientific Discovery,* edited by Lorenzo Magnani, Nancy J. Nersessian and Paul Thagard, 75–83. New York: Kluwer Academic Press.

Suárez, Marucio. 2003. "Scientific Representation: Against Similarity and Isomorphism." *International Studies in the Philosophy of Science* 17 (3): 225–244.

Suárez, Mauricio. 2004. "An Inferential Conception of Scientific Representation." *Philosophy of Science* 71: 767–779.

Suárez, M. 2010. "Scientific Representation." *Philosophy Compass* 5 (1): 91–101.

Wilder, Kelly. 2009. *Photography and Science.* London: Exposures.

7 Brains, Windows and Coordinate Systems

Annamaria Carusi and Aud Sissel Hoel

INTRODUCTION

Brain images are currently among the most important scientific images to have caught the cultural imagination. Promising to offer new insights into the brain, the mind and the self, recent and emerging neuroimaging methods have elicited an enormous amount of attention also beyond science and academia—in public media, museums and art—spawning what has been referred to as a "veritable neuro-everything craze" (Pustilnik 2009; see also Thornton 2011). An important reason for the increasing prominence of neuroimaging is the distinctive visual status of these tools. Brain images are often assimilated to photographic images and to x-rays, and to be a form of "window" onto reality as those are often claimed to be. There has been a great deal of analysis by now to show how erroneous these identifications are. As in other cases where the window metaphor is used, this conveys a simplified and misleading understanding of scientific vision and the mediating roles of scientific instruments. The window metaphor fosters flawed conceptions of photographic and medical images as transparent conduits that provide an unmediated access to reality. The resulting "myth of transparency" has been thoroughly criticized by scholars in media studies and science and technology studies (Beaulieu 2000; Sturken and Cartwright 2001; Dumit 2004; Joyce 2008). Besides that, further work has shown that the interpretation of scientific images is a learned activity (Lynch 1985; Goodwin 1994; Grasseni 2009), and that brain scans, being digital, are malleable fields (Alač 2011; de Rijcke and Beaulieu 2014).

In this chapter, instead of comparing brain images to photographic images, we bring them into juxtaposition with another form of visuality, that of painting, specifically paintings that convey spatiality in some way. In drawing this comparison, we draw inspiration from Maurice Merleau-Ponty's seminal essay, "Eye and Mind," where science and painting are also compared and contrasted. We read neuroimaging through this essay, but also use neuroimaging to interrogate the dichotomy that Merleau-Ponty draws between science and painting, arguing that neuroimaging shares some of the characteristics that he attributes to painting alone. Above all, our aim in this chapter is to contribute to the growing analyses of brain images that question the rhetoric of these images, be this in the public domain of science communication, or among scientists. One of the main problems with the myth of transparency is that it portrays, as pointed out by Kelly A. Joyce, the image and the real as interchangeable. By conceiving the image as an unmediated slice of reality, it prevents us from realizing how image and reality coconstitute each other (Joyce 2008, 75). The focus on coconstitution also shifts attention away from accounts of neuroimaging and other scientific imaging in terms of social construction. Like Morana Alač, we propose a change of emphasis, calling attention to brain images "as the center of action *with* which (not only *on* which) the work is performed" (Alač 2011, 18). Following up on this, instead of pursuing the antagonistic ideas of transparency and social construction, we aim to develop an alternative account of brain images that accentuates the generative dimension of neuroimaging methods. This includes a further development of the "with which" approach just alluded to, which in our case will be based on conceptual resources provided by the philosophy of Merleau-Ponty.

There is a further metaphor that is implicit in the rhetoric surrounding brain images, and indeed reinforced by their digitization: As Beaulieu points out in her analysis of digitally enabled neuroscience, the "god trick persists: our technoscientific investments are still partial but still claim to be total" (2001, 237). This brings to mind Haraway's critique of the "god trick" of visualizing technologies laying claim to a form of objectivity guaranteed by a "view from above, from nowhere" (1988)—a realization of the desire for a purely objective viewpoint, unloosened from any human or technological process of construction and untethered from any particular viewpoint— something that is put forward by Nagel as the aim of science (Nagel 1986). It may seem that the coordinate systems that are pervasive in computational visualizations, including in contemporary brain visualizations, are a major step toward the achievement of such a goal, since, being "object-centered" (Willats 1997), these coordinate systems appear to enact precisely a view from nowhere in particular.[1] However, in this chapter we develop an alternative approach to coordinate systems and visualizations, conceiving them as "systems of equivalences" in the terms of Merleau-Ponty.

The two metaphors, of the window and of the view from nowhere, both have implications for the understanding of the spatial and epistemic relations between science and its objects that are particularly relevant to vision and images in science. In "Eye and Mind," Merleau-Ponty puts Cartesian optics and space at the center of his critique of science, and he uses the paintings of artists such as Paul Cézanne, Henri Matisse and Paul Klee as a counterpoint to the type of science that has been erected upon a partial understanding of Descartes. It is important to note that this is not an aesthetic point that Merleau-Ponty is making (i.e., having to do with beauty and related evaluative concepts). Rather, he holds that there are forms of painting that allow us to see vision at work—the same vision that is at work in scientific practice as in painting. Our aim in this chapter is to show how this "laboring" vision, as understood by Merleau-Ponty as functioning in painting, is also at work in neuroimaging, and how it is at once formative and revealing of neurophenomena. Drawing out these "painterly" aspects of neuroimaging complements studies that have focused upon the coconstitutive nature of images and reality and the social embedding of neuro-imaging practices; but it also approaches these from a different angle that ultimately demands giving up entirely on contrasts between transparency and opacity, or between complete and partial perspectives. In order to conceptualize the simultaneously formative and revealing role of neuroimaging as mediators of the neuroscience domain, a new and integrated account of vision, images, objects and space is needed, which—as we argue—germinates in Merleau-Ponty's later philosophy of science and vision in painting.

In the next section, we shall sketch out the broad outlines of what is brought to an analysis of neuroimaging by seeing it through the lens of Merleau-Ponty's accounts of vision and painting in "Eye and Mind." Each section after that will expand on an aspect of this account: The second section sets out in more detail the account of vision in science and painting, contrasting "operationalist thinking" derived from Cartesian optics and space with painting that "shows forth" the labor of vision. The third section introduces the terms "poetic vision" and "system of equivalences," which are key to the account of vision as both formative and revealing. This section includes a discussion of lines and borders, and perspective as forms of laboring vision, and thus as a form of "poetic" or generative vision—even if it does not always announce itself as such. In the fourth section, we discuss brain images and visualizations focusing upon the way in which they are produced, used and understood to convey the spatiality of the brain. Our exploration of the spatiality of brain images—the way in which they are themselves spatial objects used to make sense of the brain as a spatial object—uses 2-D and 3-D brain atlases as examples. We show that the handling and understanding of space through these images have affinities with the process of painting with respect to the establishment of what Merleau-Ponty refers to as "systems of equivalences," and that neuroimaging as a process and in its practical use can be seen as yet another form of "laboring" or "poetic vision." Finally, we conclude with a discussion of the implications of seeing neuroimaging as a form of poetic vision for the critique of the metaphors we started with: of the window and of the view from nowhere.

Why "Eye and Mind"?

Even though Merleau-Ponty could not have imagined the forms of imaging now available to neuroscience, this essay, published in 1961,[2] challenges science specifically on the way in which it handles space and therefore vision. Turning to Merleau-Ponty in order to elucidate the visual status of current advanced neuroimaging methods may at first seem an odd choice—even more so with a view to the fact that "Eye and Mind" starts out with what appears to be a forthright attack on science: Science manipulates the world through its technologies, giving rise to artificial laboratory constructs; it treats everything as indifferent objects-in-general, reducing thinking to mere data-collecting techniques and the objects of science to mere data collections. On the face of it, today's brain sciences live up to Merleau-Ponty's worst nightmares: Disembodied brains are warped into conformity and strapped into stereotaxic grids, ready to be sliced in every direction and yield their innermost secrets. Nevertheless, the story that we draw from this essay is a more nuanced one. First, as it turns out, the "science" attacked by Merleau-Ponty is not science *in practice* but a certain way of *thinking* about science, sometimes implicit and sometimes formalized into a philosophy of science, which he refers to as "operationalism" (1993, 122–123, 137 EM). In Merleau-Ponty's view, what characterizes this kind of thinking about science is that it fails to recognize the internal complicity between vision and world. It fails, that is, to recognize the crossover between the seer and the visible, and conceives vision instead "as an operation of thought that would set up before the mind a picture or representation of the world" (1993, 124 EM). In this chapter, we want to draw attention to a similar failure to recognize the internal complicity—this time between image and reality—to be found in the kind of thinking about images that is promoted by the window metaphor. Second, in order to elaborate his account of the crossover between seer and seen, Merleau-Ponty turns to painting. Certainly, for Merleau-Ponty, all paintings are complicit in what they make visible. However, whereas some paintings endeavor to elide or suppress the traces of this complicity (he mentions the perspectival techniques of the Renaissance), other paintings (such as the paintings of Cézanne) draw explicit attention to it. The latter kind of paintings bring to the fore the process of painting, and more importantly, what Merleau-Ponty refers to as the "labor of vision"—"labor" here to be taken in the dual sense of "giving birth" and "work." By juxtaposing the process of neuroimaging with the process of painting as described by Merleau-Ponty, we aim to show that brain images have in common with paintings a complicity with what they make visible; furthermore, that there is even scope for considering neuroimaging as in a certain sense analogous with the type of painting that Merleau-Ponty describes as showing forth the labor of vision. This mode of neuroimaging undercuts the rhetoric of brain images as providing a neutral window onto the brain or mind.

The window metaphor promotes a kind of thinking that conceives unmediated presentation as the ultimate goal for visualizations of science. The development of pictorial media since the Renaissance has been characterized by many scholars in terms of a tension between transparency and opacity. This development has also been understood in terms of a tension between two different ways of seeing, that of "looking through" and "looking at" (Barthes [1980] 1981; Bryson 1983; Crary 1990; Lanham 1993;

Bolter and Grusin 1999). However, as we shall see, the internal complicity between vision and world that Merleau-Ponty is getting at in his account of embodied perception, and which is exemplified by painting in "Eye and Mind," is neither transparency nor opacity. It concerns instead the generative powers of vision and painting. This is precisely where the "with which" approach referred to earlier comes into its own. Vision and painting do not imitate the visible, like "a tracing, a copy, a second thing"; rather, they "render visible" (1993, 126, 143 EM). A key passage in "Eye and Mind" elaborates on this alternative way of approaching images:

> The animals painted on the walls of Lascaux are not there in the same way as are the fissures and limestone formations. Nor are they *elsewhere*. . . . I would be hard pressed to say *where* the painting is I am looking at. For I do not look at it as one looks at a thing, fixing it in its place. . . . Rather than seeing it, I see according to, or with it. (1993, 126 EM)

Therefore, rather than transparency or opacity, and the implicit spatial relations they evoke, the notion of seeing "according to" or "with" a painting draws attention to the roles images play in shaping both the way we see and what we see.

The metaphor of the transparent window, which implies a fixed viewpoint, is also paradoxically associated with that of the "view from nowhere" which (if it were possible) would give a view onto things as atomistically spatial—that is, as unrelated to any other thing and unrelated to any viewer. The assimilation of the projective techniques of perspective with Descartes' abstract mathematical space is now a commonplace in visual studies (Panofsky 1997; Damisch 1995), and is associated with a certain way of seeing referred to as "Cartesian perspectivalism" (Jay 1994). As we shall see, Merleau-Ponty also makes a connection between Cartesian ideal space and perspective. Interestingly, however, "Eye and Mind" proceeds to question the Cartesian interpretation of perspective and hence to challenge the established conceptions that tend to see it either as a naturalized tracing of reality or as a controlling view from above. We use Merleau-Ponty's discussion of Cartesian space and perspective as a springboard for a reflection on the role of coordinate systems in neuroimaging.

Coordinate systems as applied in the production of scientific images and visualizations seem to make something like a "view from nowhere" concretely realizable in today's computerized situations where stereotaxic grids underlie visualization systems of all kinds and are used specifically for "realistic" three-dimensional rendering of data. John Willats (1997), for example, maintains that the visualizations produced by computer systems are derived from "object-centered descriptions"—that is, from "a list of coordinates that describes the shape of an object independently of any particular point of view" (Willats 1997, 171). In this, computational visualizations are contrasted with pictures in linear perspective, which—again in Willats's terms—are derived from "viewer-centered" descriptions (ibid., 168–174). In contrast to this, other scholars such as Jay David Bolter and Richard Grusin maintain that a "logic of transparent immediacy" is at work in both the paintings of the Renaissance and today's digital 2-D and 3-D visualizations, and further that the logic of immediacy in both cases is bound up with an opposite tendency toward awareness of the medium, which they refer to as the "logic of hypermediacy" (Bolter and Grusin 1999, 20–44). In this chapter, we argue for an alignment between painting and neuroimaging. Drawing on Merleau-Ponty's discussion of vision and painting, we argue that the images and coordinate systems used by neuroscientists never function as transparent conduits; they function instead as what Merleau-Ponty refers to as a "system of equivalences." With this term, there is an attempt to understand what makes perspectives plural and mobile, but also what connects and disconnects them; what makes them commensurable or incommensurable with other perspectives, or in Merleau-Ponty's terms, "incompossible" (1993, 129 EM). Neither a fixed viewer-centered perspective nor the disembodied and indifferent shifting of an object-centered perspective allows for the kind of visibility that emerges in painting—and, as we shall argue, in neuroimaging too. Instead, the notion of system of equivalences allows us to think of perspective as stabilizing, yet never static; as formative of vision, rather than as a series of detached visual vantage points.

Science and Cartesian Space

In the first part of "Eye and Mind," Merleau-Ponty contrasts the artifice of science with the naturalness of painting. Science manipulates the world through its techniques and technologies, and gives rise to constructed laboratory objects. This is not, however, the reason that Merleau-Ponty criticizes science. His criticism is of a mistake that science makes when it takes the constructed laboratory object for the naturally occurring object, a mistake that amounts to a misunderstanding of itself and its own processes, as much as it is about the objects studied by science. The attitude behind this mistake is one that Merleau-Ponty traces back to a philosophy of science instituted through a certain partial reading of Descartes. Through pursuing a line of thought derived from Descartes, science forgets or puts in the blind spot the "primordial historicity" of the body and objects, which precede and exceed the laboratory object. What is at stake in this forgetfulness is a proper understanding of scientific vision, or, more broadly, the way in which science gains access to its world.

Merleau-Ponty's fascination with painting and art in general was motivated by his belief that these are important means for revealing the nature of vision and signification in general. In "Eye and Mind," he refers to painting as being a "secret science" of vision, which stands in contrast to vision understood scientifically. The latter he believes to be molded by a way of thinking about vision and space epitomized by Descartes. We start with a brief summary of what for Merleau-Ponty is an "operationalist" account of vision, and then go on to the "secret science" of vision that painting is.

It is through Descartes' few comments on painting that Merleau-Ponty introduces the Cartesian notion of space, as he holds that even though Descartes does not often talk about paintings, when he does, he reveals more of how he thinks about vision and space than what is made explicit in the *Optics* (Descartes [1637] 1965).[3] For Descartes, depth, or the third dimension, is derived, rather than experienced, from the other two dimensions, height and width, acting like signs, to give the *illusion* of depth. Occlusion of one thing by another is one of the most important signs that paintings use to convey the illusion of depth. Depth as conveyed in painting is an illusion because in real extended space, which is accessible only from a god's eye view, nothing is in front of or behind anything else; things are outside one another, always in full and complete view. For Descartes, as extended beings, we *are* in space, but we can truly *know* space only when we abstract our bodily being from it; we know space only through reason and hence as an ideal mathematical space.[4] This externality of an embodied viewpoint to space is captured in the often misleading analogy between perspectival paintings and windows, which implies that the two-dimensional surface of the painting opens onto depth as it would exist for bodiless beings. Thus understood, Merleau-Ponty observes, "the window opens only upon *partes extra partes*, upon height and breadth merely seen from another angle— upon the absolute positivity of Being" (1993, 134 EM).

Merleau-Ponty's critique of science needs to be contextualized in his discussion of Descartes' conception of space. He does not in fact criticize Descartes for having idealized space, for having given an account of it that is not empirically informed—nor for having proposed a notion of space that is a construct. On the contrary, he sees this stage of idealization and construction as a necessary step in the "freeing" of space, allowing for a different thinking about it than what appears to be accessible empirically (empiricism can be blind to its presuppositions too). Rather his criticism is targeted only at one point—namely, where this construction is erected into a positivity—that is, when the construct of space is treated as though it simply were space as such, the only "real" space. In fact, things are more nuanced even for Descartes, whose treatment of space splits off in two directions, one for which space is a construct that is "clear, manageable and homogenous" and can be fully known, and another for which space is a "blind but irreducible experience," tied to our bodily immersion in it (1993, 134, 137 EM). What characterizes "operationalist" science, then, is that it dispenses with the second of these notions, appropriates the first as an exhaustive notion of all kinds of space and assumes the complete certainty of its own contact with the world. This is an attitude of forgetfulness on the side of science, the cost of which is ignorance of its own processes and formative activity. This forgetfulness is something that science has in common with what Merleau-Ponty refers to as "prosaic vision," to which we return ahead.

It is important first to discuss one far more explicit legacy of Descartes in the realms of science, and that is the Cartesian coordinate system, which is a certain way of putting into practice his idea of space as a logical construct, the use of which is widespread in science. With this system, Descartes provided a way of visualizing abstract symbolic algebra as spatial geometry, starting with the association of the variables in algebraic formulae with the two dimensions of a plane (the axes x and y, adding depth as a third dimension, z), and of the values with points along the axes, resulting in a spatial plotting of the solutions of algebraic equations. With this simple yet powerful mechanism, Descartes bridged the gap between algebra and geometry, and, more importantly for us in this chapter, invented one of the earliest forms of visualizing information. This single innovation has been profoundly formative of scientific visualization and still is (van Fraassen 2008, 22). Indeed, Cartesian coordinates, complexified far beyond the initial two or three axes of equations with only two or three variables, are still very much used in computational visualizations of all kinds.

It is this technologization of Cartesian coordinate systems that appears to warrant the rhetoric about the subject-free objectivity of these visualizations. However, as we show in this chapter, neuroimaging in its use and context does not fit this rhetoric. Coordinate systems as they are used to convey the spatial features of neurophenomena have more affinity with the way painting deals with space than with idealized notions of space. In fact, we argue that neuroimaging has formative capacities and that it is a form of "poetic vision," as we go on to discuss in the following sections.

PAINTING AND THE LABOR OF VISION

The conception of space that assumes things to exist *partes extra partes* is also referred to by Merleau-Ponty as "prosaic space" and corresponds to "prosaic vision" (or sometimes "profane vision"). This a conception of vision that assumes the externality of vision to what is seen, as if the visible world were constituted independently of any particular mode of seeing. Vision is assumed merely to represent what is already there, as if the objects of the world, such as apples and meadows, were already parceled up for vision. The notion of picturing that corresponds with prosaic vision proceeds as if objects were given with lines already drawn around them, which pictorial depiction need merely reproduce (1993, 142 EM). Forgetful of its own stake in forming vision, prosaic vision accepts these lines as positive features of the objects themselves, existing independently of vision. In this way, "the visible in profane vision forgets its premises" (1993, 128 EM).

Merleau-Ponty turns to forms of painting that do not accept this atti-tude, but which explicitly recognize that there are no preexistent lines or contours to identify and differentiate objects. Lines and contours are never strictly in any identifiable place—"they are always on the near or the far side of the point where we look"; "between or behind"; they are never things in themselves (1993, 143 EM). These paintings show vision to be poetic—so-called not because it is a form of aesthetic vision, but because it is formative and generative. Poetic vision refuses the separation between itself and what is seen. It attends instead to the means through which things become visible in the first place, in a process in which both are complicit. By "complicit" is meant, first, a mutuality between seeing and seen, and second, that the process is hidden, not obvious and that it needs to be revealed (hence the term "secret science" of vision).

The treatment of lines and borders, which Merleau-Ponty already dis-cussed in "Cézanne's Doubt" (1993, 59–75), is one way in which poetic vision is manifested. One of the examples on which Merleau-Ponty draws in this essay are certain still lifes by Cézanne, where fruits and other objects do have outlines, for without these they would be deprived of their identity. The outlines, however, are not made as one continuous line encircling the object; rather there are several outlines, and none is continuous, allowing the shapes to emerge by way of the lines that themselves recede from view. For example, if we look at *Still Life with Water Jug* (1892–1893),[5] the out-lines of objects such as the fruits consist of multiple strokes, and the way in which lines recede from view is clearly seen in the outline of the jug, giving it depth and voluminosity against the darker background. Thus, instead of making an object of the shape, Cézanne treats the contour as an "ideal limit toward which the sides of the apple recede in depth" (1993, 65 CD).[6] The lines in these instances are not imitations of preexisting lines; they are not representations; rather they "render visible" as Merleau-Ponty puts it, quoting Paul Klee (1993, 143 EM).

It is for example through these treatments of line that painting reveals something about embodied vision. Looking at this painting by Cézanne, we cannot take vision for granted; rather it reveals vision at work. It shows that the premises of vision are material, positioning seers and seen in rela-tion to each other and not in some abstract space. It also shows vision to be processual, and not there "all at once." So, for Merleau-Ponty, the forma-tive exchange between objects and lines iterates and extends the formative exchange between embodied vision and the world, for in both cases vision does not merely copy or reproduce; there is a *process* of seeing, and that means a *way*, *mode* or *style* of seeing, among others. The point of these paintings is not that there is a resemblance between their inherent features and the features of a pregiven object, but rather that they organize and structure the way that we see and access objects in the world—for example, even in the way that we pick out some features rather than others. The point of these paintings, in other words, is to emphasize the generative dimension of vision, which is overlooked by Cartesian and representational notions of vision and picturing. The work performed by the lines in Cézanne's paintings is one aspect of what it means to see "according to" or "with" a painting, rather than just seeing it. This is what we were alluding to with the "with which" approach referred to in the introductory section.

In the literature on images, linear perspective often features as the representational mode of depiction *par excellence*, as an instantiation of Cartesian vision and the epitome of a detached view from above. Merleau-Ponty also alludes to this interpretation when he sets out to explain why Descartes exemplifies his writings on vision with line drawings rather than with colored paintings. For Descartes, it is line drawing that makes painting possible, preserving the form of objects and allowing the representation of extension. On this view, drawings in perspective are governed by a projective relationship between the picture plane and objective space, the rules of which, it was believed, had been successfully mathematically formalized. However, colors are not governed by rules and ordered correspondences. So, if Descartes had attended more to color, Merleau-Ponty conjectures, he would have been obliged to give up the prosaic conceptions of vision and space, as well as the prosaic notion of the line as imitating the preformed shapes of things. For, just as the idea of laboring vision goes counter to the conception of vision as "an operation of thought that would set up before the mind a picture or a representation of the world" (1993, 124 EM), the poetic notion of lines and colors goes counter to the conception of painting as "an artifice that puts before our eyes a projection similar to the one things themselves would (and do, according to the commonsense view) inscribe in them" (1993, 133 EM). Merleau-Ponty is not here delivering the familiar critique of perspective as Cartesian; rather he aims to include perspective in poetic vision, and thus "to integrate perspective, as a particular case, into a broader ontological power" (1993, 133 EM).[7] Practicing painters, Merleau-Ponty maintains, have always known that perspective is not an exact and infallible way of conveying spatiality—nor is it the only way. The method of perspective is not instituted by nature; rather, it is "only one particular case, a date, a moment in a poetic information of the world which continues after it" (1993, 135 EM). Thus, for Merleau-Ponty, all paintings are instances of poetic vision, but some kinds of painting call attention to their own formative capacities in ways that others do not.

So what are the implications of this when it comes to understanding the ways spatiality is conveyed in perspectival paintings? Again, Cézanne is a good example, this time in his profound interrogation of space in the series of paintings of the Mont Saint Victoire.[8] In these paintings, the mountain is painted from multiple perspectives at once, so that there are different perspectives from different points of the picture plane. The spatiality of the scene, the depth and volume of the mountain, emerges out of the interrelatedness and mutual confrontation of these perspectives. Each painting embodies a multiplicity of perspectives, which from a single viewing point in front of the paintings can give an impression of disjointedness.

As viewers move in front the paintings, the transition from one perspective to another becomes evident, and with that, the impression of disjointedness gives way to a shifting mobile perceptual experience of depth and solidity. The voluminosity of the mountain, in other words, emerges out of the tensions between these incompossible perspectives. Again, the painting reminds us of the labor of vision, which plays itself out in space and time. This means that *even a single perspective is already multiple* in that it is already implicated in a series of alternative perspectives. Vision is always beyond the here and now. In this way, Merleau-Ponty argues that even classical perspective is poetic, even if it is often mistaken as an external and passive viewpoint mimicking objects and scenes. Perspective is poetic because it always involves a formative intervention of vision, even when it does not make this manifest.

The reason Merleau-Ponty turns to the paintings of Cézanne is that they interrogate and unveil the structures that "exist only at the threshold of profane vision" and that "are not ordinarily seen" (1993, 128 EM). As we have seen in the previous two sections, vision involves formative structures that make vision possible even if they go unnoticed. For Merleau-Ponty, these make up what he calls "systems of equivalences." From his earliest work, the notion of system of equivalences is associated with the notion of "style," since for Merleau-Ponty "perception already stylizes."[9] Style submits the elements of the world to a "coherent deformation" through which phenomena take on shape and meaning. It is a *deformation* because it never passively imitates but, in a way, warps the elements of the world into a meaningful environment. This deformation, however, is not a distortion but an orientation that is productive of an environment with specific features and arrangements. The *coherence* of the deformation lies in the patterns of what becomes salient in that specific environment. Thus, on this view, a style is a system for making figure-ground patterns (1993, 91 ILVS). One aspect of this coherent deformation is the way in which perception is always discerning a figure against a background that can obviously vary between people in different circumstances. As we have already seen, this discernment also takes place in painting, where lines actively bring out the shapes of things and thereby make them visible against a less specified background. As Merleau-Ponty makes clear, this discernment even takes place in linear perspective, which is not simply a reproduction of a pregiven space, but displaces previous spatial orders by generating new systematic arrangements of objects. In these spaces, phenomena become articulated and hence identifiable and comparable along certain dimensions specified by the style in question. A style is a system of equivalences, then, in that it serves as a kind of *yardstick* according to which phenomena are rendered comparable without being collapsed into sameness.

That perception already stylizes means for Merleau-Ponty that already the perceiving body is a system of equivalences. As we already discussed in the case of painting, vision does not simply impose an order but is complicit with objects; a system of equivalences is the means through which this complicity is accomplished. Thus, in embodied perception, the seer and the seen enter into a "strange system of exchanges" (1993, 125 EM). Beyond that, as Merleau-Ponty makes clear in *The Visible and the Invisible*, this system of exchanges also happens across sense modalities. Vision and touch, the tangible and the visible, interimplicate each other; they form, so to speak, two maps that align without ever merging into one (Merleau-Ponty [1964] 1968, 134). These exchanges are also accomplished by style as it is found in art. For example, the same style can be recognizable in works by one artist in different modalities, such as painting and sculpture. According to Merleau-Ponty, this is attributable to a system of equivalences—that is, "a Logos of lines, of lighting, of colors, of reliefs, of masses" (1993, 142 EM) that stretches across the modalities in which an artist works. Hence, when it comes to understanding the individual styles of painters, what makes "a Vermeer" is not so much that it was painted by Jan Vermeer the historical person, but that the painting observes a system of equivalences that marks the "same deviation" as other Vermeer paintings (1993, 98 ILVS). At this point, then, we can say what it means to "see according to" an image such as a painting: It is to adopt the system of equivalences of the image. This takes us beyond the dichotomy of transparency and opacity, since an image is not something that we see *through* or look *at*; rather an image is a system of equivalences *with which* we see. Images, therefore, never allow a view from nowhere; as systems of equivalences they are oriented toward something and from somewhere. This is what style does: It hinges together these two orientations that prosaic vision tries to keep apart. Similarly, we need to move beyond the dichotomy that opposes single complete perspectives to multiple partial perspectives. The crucial point is not that there are in fact multiple perspectives and no single correct or objective perspective, be it a view "from nowhere" or "from above"; rather every perspective, even the most stringent classical linear perspective, is poetic in that it is a system of equivalences that stands in a generative relation to what it targets. It is as a system of equivalences that vision works, and it is as a system of equivalences that painting becomes a labor of manifestation (1993, 91 ILVS).

We had set out to show how Merleau-Ponty's discussion of painting is relevant to neuroimaging. In the next section we will show that, despite the predominance of prosaic conceptions of vision in the discourse of imaging and visualization, neuroimaging is a form of poetic vision in the way in which it conveys the spatiality of the brain. In the next section, we will draw attention to the "painterly" aspects of neuroimaging by showing how systems of equivalences are at work in them.

THE "PAINTERLY" ASPECTS OF NEUROIMAGING

In "Eye and Mind," Merleau-Ponty draws a distinction between science and painting, seemingly attributing prosaic vision to the first and poetic vision to the second. It is important to note that "vision" for Merleau-Ponty is taken broadly, including not only the formative power of the body but also the formative powers of technologies of vision and observation (Carusi and Hoel 2014). For example, the construction of laboratory objects clearly involves more than bodily vision, and it is not the practice of this construction that is his main target. Rather, as we have already pointed out, his criticism is of a certain way of thinking about science that is forgetful of the labor of vision in that it does not acknowledge the formative contribution of vision in constituting its objects. Even though it might appear that this criticism would apply to neuroimaging too, our analysis will show that neuroimaging is poetic in similar ways to painting, and that these formative aspects are not as deeply hidden as might be supposed. In neuroimaging, knowing where you are in the brain is paramount, but comprehending the spatiality of the brain is very challenging. The neuroscientists observed for this chapter are acutely aware of this as well as of their own role and the role of technologies in making the spatial aspects of their object accessible.

The discussion of this section is based on fieldwork conducted by both authors in two neurolabs. One lab (Lab 1) conducts research primarily on the plasticity of the visual system, using both human and animal models. The other lab (Lab 2) conducts research on functional neuroanatomy of networks in the brain that mediate learning and memory, using mostly rodent models. Both labs use a wide array of imaging techniques and experimental methods. Lab 1 uses mainly 2-D atlases, both paper and web-based atlases, and has also contributed to developing an online atlas consisting of serial histological sections of monkey brains. In addition to using and producing atlas devices, Lab 2 is involved in developing computational tools for visualizing the brain, such as an interactive diagram of connectivity. It is also collaborating in developing a hippocampus atlas (a "virtual microscope" that allows for high-definition zooming and juxtapositions of sections made through different staining techniques), and an atlas of the rat brain combining 2-D and 3-D views.

Our fieldwork has included participant observation, interviews with the lead scientists of the two labs, attendance of seminars and lectures, and observation of user groups. One of the authors (Hoel) also observed brain surgery. Apart from the two labs, we also interviewed a senior researcher in medical imaging about the use of coordinate systems in his field.

In all neuroimaging spatiality is key. A clear example of this is image-guided neurosurgery, where navigation accuracy depends upon an alignment of coordinate systems of physical space and image space. Through the process of registration, images (for example, preoperative MR) are transformed into the coordinate system of the patient or aligned with images of a different modality (for example, intraoperative ultrasound) (Lindseth et al. 2013). In fact, the Talairach coordinate system, which for a long time was the gold standard for coordinate systems in neuroimaging, was initially developed for neurosurgery and stereotactic surgery. The way in which the spatiality of current images is dealt with can be traced back to the first attempts at the end of the nineteenth century, and the first stereotactic frames based on a coordinate system at the beginning of the twentieth (Enchev 2009). The technological development of coordinate systems has played a crucial role in the constitution of neuroscience, and today these systems are an in-built part of the technologies that define this science. As we have already mentioned, Descartes initiated this way of conveying mathematical structures through coordinates, which today are embodied in imaging and visualization systems.

Spatial accuracy is important for precise guidance, orientation and localization in neurosurgery, and, as is well known, for mapping cognitive functions onto the physiological brain in cognitive neuroscience. Functional neuroanatomists investigate structure-function couplings at the level of the brain as organ and at the level of relationships among brain cells. Hence, knowledge about regions, areas and connectivity between them is an intrinsic part of their experiments, as is the accurate spatial localization of their interventions. However, it became clear in our encounters with neuroscientists that even with the aid of coordinate systems, understanding the spatiality of the brain is by no means easy. The difficulties are due in large part to all the different spatialities involved. For example, in the different modalities of neuroimaging scientists have to deal with the brain in its own space, the brain in the head and the brain in the physical world, as well as brain templates and average brains, or different combinations of these depending on the task at hand. In each of these mediated situations, there is a disruption of embodiment relationships between viewer, image and brain. In our fieldwork this came out in the way that practitioners are constantly trying to understand the orientation of the target brain and simultaneously to position themselves vis-à-vis their visualizations. The most common phrase that peppers the discourse of these neuroscientists relates to "where you are," as in "knowing where you are," "you think you are there, but you are not," "mapping where you are" and also "if you are lost in the brain." In all these cases, "you" refers to the scientist: that is, the spatial orientation in the image is grasped in an embodied way, either as though the scientist is located within the image, trying to make sense of it as of a territory in which s/he stands (for example, at the insula, at the basal ganglia), or as though the scientist is standing before an embodied brain. An example from one of our informants when trying to make sense of the orientation of a 3-D brain is the statement: "This brain is looking at you" (Session 2 transcript).

The point we want to make here, however, is not simply that the disruption of embodiment relationships is a distortion that needs to be rectified through re-embodiment. Rather, in our account of scientific vision, the disruption is productive or generative in the sense that it *allows scientists to see more.* Noting that scientists locate themselves with respect to the image through immersing themselves imaginatively in the picture space and using bodily gestures goes some way to account for how they deal with spatiality. However, in order to grasp the generative dimension of scientific vision, a further point needs to be made that emphasizes even more pronouncedly the active role of vision in discerning the objects under study. We will now turn to some examples from our fieldwork that show the labor of vision at work in understanding the spatiality of brain images.

Our first examples relate to the delineation of borders around areas and regions in the brain. The first example taken from our fieldwork in Lab 1 concerns the interpretation of images obtained through fMRI experiments on vision. In some cases, the images are acquired from specific subjects in one specific scanning session. The images are interpreted by relating together the coordinate system, brain atlases and knowledge of structure and expected activation patterns. There can be different coordinate systems at play: One is the coordinate system of the software for image capture and analysis, and another is a coordinate system that researchers will themselves impose on an image for purposes of orientation. For example, when shown for the first time an online atlas that makes the axial lines explicit and allows users to choose their own point zero, our informants instantaneously recognized a practice that they carry out for themselves all the time in their lab. A major point of orientation is point zero in the coordinate system—that is, where the three axes (x, y, z) intersect. This point could either be given conventionally (for example, according to the Talairach system), or else be a point commonly used by the group. For example, our informants use the intersection between the "interaural" line when looking at images of animal brains—that is, the imaginary line running from ear to ear—and for the other axes, a landmark that is easy for them to recognize, or a point on the skull derived from the crossing of the coronal and sagittal sutures. In these cases, the landmarks serve as points of reference. Even if the axial lines are in a sense imaginary, they are also shared among the group through their experimental practices (for example, placing landmarks on the skull during procedures), and in the analysis, through pointing at features of the image. Thus, even when there is not a material or visible stereotactic frame in an image, the scientists will enact one, and go back and forth between their interpretation and the reference brain atlas. Through this enactment, the spatial landscape of the brain area in question emerges.

When they are looking at an fMRI image, our informants will look at where activations occur, and consider their values. For example, if they are at 85 (a voxel number assigned in the software coordinate system), they will assume they are "going toward the occipital lobe." However, they cannot be entirely sure of this, since the coordinates of the image are not absolute, but relative to the specific experimental setup, including the scanner, the software used, the position and size of the head in the scanner, and individual differences in the brain. If they get a value for an activation lying outside of the region they would expect, they are not sure to what factors it could be attributed: Could the subject have moved? Could the registration be incorrect? At this point, they will check the reading against an atlas. In conversation, they likened this process to that of driving a car to some unknown point, looking at a landmark such as a street sign (that is, the value obtained) and locating that landmark on the map, so that you can see what region you are in. In this process of locating activations, the borders of the regions for any image do not clearly preexist the interpretation; rather they emerge through the working together of the scans, the coordinates provided by the software, alternative coordinate systems that they use in their practice, knowledge of anatomical landmarks, and the reference brain atlas. The ability to draw the borders for themselves is very important—for example, when shown examples of 3-D images where the boundaries are demarcated by using different colors for different regions, one of our informants said, "We don't use those precolored things." This resonates with the treatment of lines and borders in the paintings discussed in the previous section, where lines do not merely encircle pregiven objects but instead borders are rendered visible through the working together of several modes of drawing lines.

The issue of borders was even more pronounced in Lab 2, where common mistakes of localization were even attributed to an overreliance on the stereotaxic coordinate system. One of the sessions we observed dealt with subdivisions of the hippocampus and nearby areas in rodent brains. In presenting a hippocampus atlas (the online virtual microscope already referred to) developed by his lab, the scientist leading the session pointed out that using only the coordinate system can lead to mistakes about which areas one is recording during experiments. The coordinate systems must be supplemented by another visual method—that is, cytoarchitectonics (the study of cell bodies and composition under a microscope and the method of parceling the brain through stained sections of brain tissue). The hippocampus atlas developed by Lab 2 includes general cytoarchitectonic descriptions of the various hippocampal and parahippocampal areas as well as descriptions of the borders between these areas. For, as pointed out by our informant, sometimes researchers claim to be in areas where they cannot possibly be. It is often the case that researchers determine what areas they are in by relying merely on the corresponding sections in standardized atlases. However, in some cases the cytoarchitectonic organization of the various areas tells you that you are in the wrong area even if the coordinates fit. As our informant put it:

And they say, I lower my cannula and inject something, but they never look at the section, they just look at the outline of the section, and so they say, "Where should we be?." They go into Paxinos, find the corresponding section, and say, "Well, in terms of space (gesturing), this should be in this area, because Paxinos tells me, this is this area." But if Paxinos makes a mistake in the atlas, or if you have a tilted section, you may end up in a very different area. So the only way to do these locations is by actually looking at your section and putting in the borders yourself. (Session 5 transcript)

Thus, as underscored by our informant, the best way to do these delineations "is by actually looking at your section and putting the borders yourself" (Session 5 transcript). In fact, one of the motivations of the lab for developing this visualization tool is to enable users to decide for themselves when they are moving from one area to the next by scrutinizing their own experimental sections.

In both the labs that we studied there is an enactment of spatial boundaries, and the scientists themselves are aware of this. In Lab 2 the awareness is extended to the development of tools to support the active intervention of scientists in drawing lines for themselves. Relying on predrawn lines may lead to errors in localization, which we have seen has serious consequences in neuroscience. Through their own drawing of lines and visual inspection, the borders between the areas emerge for the scientists, and in this active process they gain a deeper understanding of the spatial features of the areas of the brain they are concerned with. What we have here is a dynamic interplay between different ways of looking in order to orient oneself spatially, through numeric coordinates and through the qualitative visual means of microscopy. In the qualitative process of cytoarchitectonic parceling, the defining composition of the various areas is not simply pregiven but emerges through active processes of staining, slicing and imaging. In fact, staining is exemplary of the complicity of seer and seen discussed earlier, since the patterns distinguished are dependent on what can be made visible through the techniques and materials used. Therefore, the composition of the cells as revealed in microscopy is not simply a mirroring of reality parceled out in advance.

When asked directly by us about how they see the spatial volume of the brain through their images, our informants in Lab 1 answered that they primarily work with 2-D images (both microscopy and MR images) in sequence, from one end of the brain to the other. The size of the section in the image relative to its neighbors, and to the whole sequence, gives them a sense of the volume of the brain, which they "mentally reconstruct" (in their own terms). Interestingly, when it comes to digital 3-D visualization tools, they seemed to disagree among themselves, one of them referring disparagingly to current use of 3-D visualization tools as "just for having sexy images," while his colleague saw them as continuous with other, more traditional teaching tools, such as 3-D material models of the brain and learning through dissecting. Scientists in Lab 2, for their part, are collaborating in building a 3-D tool that is still in the development stage. In our observation of a session in which the tool was being introduced and demonstrated, use was made of displays where the screen is divided into four planes: In three of the planes, there are 2-D images, and in the fourth there is a 3-D image showing the position in the brain volume of the corresponding 2-D images. The session proceeded with the lead scientist explaining the anatomical features of the hippocampal formation by pointing, by focusing on one plane at a time or by flipping between 2-D and 3-D views.

In fact, though, there is always a lot of to-ing and fro-ing between images in different modalities and between different modes of access to the brain as a volume: Understanding space through images involves not just looking at images but also undertaking processes such as handling physical models of brains, dissection, experimental procedures on actual brains such as injecting stains, positioning electrodes and so on. There are also different perceptual modalities involved, including touch, as well as a wide range of practical actions taken, including measuring, cutting, slicing, drawing, etc. It is only through the interplay between modalities and actions, where none is isolated, that researchers know where they are in the brain. The brain as a volume, in other words, is always multimodal. In neuroscience as elsewhere, understanding space through images entails seeing more than what is actually "in" the images, for, just as in embodied perception, vision is more than what meets the eye. We saw this in our discussion of painting as well; even when looking according to 2-D images, vision is already in a sense 3-D.

Traditionally, dissection has played a very important role in training researchers in neuroanatomy, but as confirmed by our informants in both labs, this is currently receding as it is more difficult to obtain brains, especially human and other primate brains. Computerization is increasingly turned to in order to make up for the experience of dissection that students would have had in the past. Digital and interactive 2-D and 3-D atlases, and the databases that provide input to them, are part of this advancing computerization of neuroscientific practice. As we have seen, scientists in both the labs that we studied are aware of the work that it takes to locate experimental data accurately in the spatial volume of the brain. For those in Lab 2 who are developing a database and tools to navigate and interact with their atlas, and collaborating on the construction of the 3-D atlas, this awareness of process becomes even more marked. The active working together of different kinds of images, coordinate systems and other experiences of handling brains is brought to the fore during the process of building the digital tools. This is also evident in the use of the tools and the functionalities of the interface, as users draw their own lines to delineate borders, work with both coordinate systems and different kinds of images, and constantly compare and contrast different modalities and visualizations. Trainees in the labs are encouraged actively to interrogate the images they use. Digital visualization, our informant explained, "allows you to image or visualize where structures are, from different angles" (Session 5 transcript), bringing out the shifting and multiplicity of perspectives, and the active relating together of perspectives one to the other that is necessary in order to see spatiality. The tools and the screen interface make up a space where different images in different modalities can be juxtaposed, easily switched between and zoomed in and out of.

The shifting between perspectives and modalities allows a comparability that is not just given but also engendered through the images and tools, and combinations thereof. In neuroscience, comprehending the spatial features of the phenomena under study always occurs through a process of inter-relating many different ways of looking. The patterns of these interrelationships is what Merleau-Ponty refers to as a "logos" or system of comparison that stretches across modalities—a "style" that establishes a certain manner of accessing and relating to phenomena. As in the case of painting, style is a marker that a system of equivalences is at work in these different ways of looking. And painting showed us that where there is style, there is a complicity between vision and its objects. This means that far from being a transparent conduit for neurophenomena, neuroimaging sets up a space of comparison that at once generates and reveals the entities under scrutiny.

When it comes to acknowledging the "labor" involved in scientific vision, and in our case, the generative dimension of neuroimaging, it is more instructive to turn to the process of developing these tools, because if we attend only to visualizations as finished products, the processes that are formative of vision and its objects in particular contexts are easily black-boxed.[10] This is in itself a form of forgetfulness that continues to feed the idea of transparency. As we saw in the case of linear perspective, even when visualizations seem to be "prosaic" they are in fact "poetic" even if they are not explicitly so. Further, in our fieldwork we found that, even if computerization is often associated with black-boxing, engaging in the development of visualization tools can serve to tease out the labor of vision, in ways that then make it more available to users. In fact, the processes of computerization that we have followed turned out to be poetic in both the senses we saw in the context of painting: They establish systems of equivalences, and they show forth the labor of vision.

Conclusion: Seeing According to Systems of Equivalences

Neuroimaging is the key technology for current neuroscience, and for neuroimaging, spatiality is the key problem. Painting has a very long history of grappling with space. In this chapter we have offered a reading of neuroimaging through the way that vision, images, objects and space are handled in painting, drawing upon the account given by Merleau-Ponty in "Eye and Mind." In paintings, lines and perspective play an active role in discerning and parceling out for vision objects with depth and volume. By juxtaposing painting to neuroimaging, we were able to draw out "painterly" aspects that play the same role for vision in neuroscience with respect to discerning and parceling out regions of the brain and other features of interest. That aspects of imaging play this active role implies that images are not simply something that you look *at* or look *through*—images are something that you *see according to*, something according to which your vision is being formed. This means that images have their own formative capacity, in the sense that they never merely mirror the pregiven but make a real difference to what is discerned and how. Nor do images simply construct their objects by imposing an arbitrary order upon them. As Merleau-Ponty insists, imaging involves *complicity* between vision and its objects; through the specific arrangements of vision, technologies and settings, objects *emerge* in different but systematic ways. We saw how this complicity plays out in the process of locating data in specific brain regions through neuroimaging. It is against this background, then, that we can say that neuroimaging is at once formative and revealing of neurophenomena.

As stated in the introduction, conceptualizing the simultaneously formative and revealing dimension of neuroimaging requires a new and integrated account of vision. By drawing upon Merleau-Ponty's philosophy of vision and painting, we approach neuroimaging from a new angle that no longer depends on the contrast between transparency and opacity, and that does not, like the inheritors of the Cartesian account of vision, set vision on the one side and objects on the other. In fact, on a Cartesian account, it would be contradictory, even impossible, to hold that vision is at once formative and revealing.[11] The reason this is possible in Merleau-Ponty's account is that he conceives of vision in terms of systems of equivalences that stretch across domains that are kept apart in the Cartesian worldview. As we have seen, Merleau-Ponty explicates systems of equivalences through the notion of style, which for him is the key to understanding both perception and painting. To say that images such as brain images are not looked *at* or looked *through* but rather that we see *according to* them means that we take on their system of equivalences, and make them part of our style of seeing. Now we can further specify what it means for images to have formative capacities on their own: This means that in order for images to work, to come to life, their systems of equivalences need to become hooked up with those of embodied viewers. To see according to images is to be hooked up in this way—something that could further account for the common style of seeing in scientific groups.

It should be clear that *seeing according to* is entirely different from the account of vision implied by the window metaphor, which essentially retains the categories of Cartesian vision, as does the metaphor of the view from nowhere. Even though object-centered accounts seem to encourage an understanding of vision that gets beyond construing images as subjective constructs, the approach we are proposing goes beyond the subject-object dualism still assumed by the object-centered accounts, by having subject and object entering into a new constellation where they are no longer opposites but part of an integrated system. This has implications for how we are to understand the rhetoric of neuroimaging. Against this background, we can now attribute the myth of transparency to the forgetfulness that is characteristic of prosaic vision, the conception of vision that is ignorant of or simply chooses to ignore the formative and generative dimension of seeing. Apart from being a source of much confusion when it comes to understanding the evidential roles of neuroimaging, this forgetfulness is also the source of the powerful rhetorical use to which brain images are often put in the public sphere. Here they are often presented as though they are perfectly complete, seeming to present a direct and exhaustive view of objects with no active involvement of vision and its technologies. Accounts such as Willats's, which hold that object-centered descriptions are independent of any point of view, manifest this forgetfulness. Still, the remedy would not simply be to swing back to viewer-centered descriptions, since this is to remain in the grip of subject-object dualism. The notion of *seeing according to* offers a way out of this binary.

This implies that the "logic of transparent immediacy" as discussed by Bolter and Grusin (1999) is not so much a logic (no image has ever been immediate) as a certain rhetoric (promoting forgetfulness about the formative powers of images). Yet, again, the remedy would not be to swing to a logic of opacity, nor to see it, as do Bolter and Grusin, as a tension between immediacy and hypermediacy. Certainly, some images call attention to their own formative capacities in ways that other images do not (in the terms of Bolter and Grusin, they raise awareness of the medium), but this should not lead us into thinking that images that do *not* raise such awareness are immediate. As systems of equivalences, all images are "mediate," but not in the familiar, representational sense of being mediate (where image and world are conceived as interchangeable), but in the very specific sense that is conveyed by Merleau-Ponty's notion of system of equivalences.

Strategies such as multiplying partial perspectives (Haraway 1988) are an important step toward breaking the thrall of the "view from nowhere," but our account makes a slightly different point. Just as in painting, even the most rigorously singular linear perspective is a mode of organizing and arranging vision, images, objects and space according to a system of equivalences that it institutes. Thus it is not multiplicity as such that renders vision productive. Images always have a generative dimension, as our discussion of linear perspective showed. However, the situation in neuroimaging is closer to the kinds of painting that Merleau-Ponty devotes most of his attention to, such as Cézanne's. As we saw in the paintings of Mont Saint Victoire, it is not just that perspectives are multiple, but also that perspectives that would normally be considered incompossible (not jointly possible) come to be interconnected through a system of equivalences, which does not so much dissolve the tensions between them as make them productive in relation to each other. This is also what we see with the many different modes and modalities of images in neuroscience. To take just one example, we saw that there is not in fact a single coordinate system at play in the neuroimaging domain, but neither are there simply multiple coordinate systems; rather there is an interplay of coordinate systems, where each is interleaved with the other without ever completely converging. In the space of comparison set up in and through these interleaved systems, heterogeneity is neither reduced nor does it lead to incommensurability that would not allow them to be brought into relation at all. Rather, by being brought into systems of equivalences, heterogeneity is made to be productive.

As systems of equivalences, images are no longer external intermediaries mediating between pregiven subjects and objects; instead they are productive interventions from which new styles of vision—and hence, new complexes of perceptual beings and entities—emerge. This means that images have formative capacities that are actualized when their style is taken up and they enter into a circuit with embodied viewers. To consider images in this way means to look for systems of equivalences that work as yardsticks that delineate phenomena at the same stroke as they locate them in a space of comparison. Thus, by providing an integrated account of vision, the notion of *seeing according to a system of equivalences* offers a conceptual and analytical tool that opens a new line of inquiry into scientific vision.

So, what we learn from looking for the painterly aspects of neuroimaging is that neuroscientists do not look at or through images—they see according to, or with them.

ACKNOWLEDGMENTS

We would like to thank our informants at both the neuroscience labs, as well as in the other contexts where we have interviewed scientists and participated in activities with them. We would also like to thank Timothy Webmoor and Liv Hausken for their helpful suggestions and comments. This chapter forms part of the interdisciplinary project *Picturing the Brain: Perspectives on Neuroimaging* (2010–2014), which is funded by the Research Council of Norway.

NOTES

1. See van Fraassen (2008, 69–72) for a critique of the attribution of a "view from nowhere" to Cartesian coordinates.
2. "Eye and Mind" (published originally in 1961) can be seen as the culmination of a preoccupation with ontology, movement and spatiality that Merleau-Ponty began to develop immediately after finishing his seminal work *Phenomenology of Perception* (published originally in 1945). While we draw primarily from "Eye and Mind" (henceforth referred to as "EM"), we also draw upon his other key essays on painting, "Cézanne's Doubt" (CD) and "Indirect Language and the Voices of Silence" (ILVS). All three of these essays are published in *The Merleau-Ponty Aesthetics Reader: Philosophy and Painting* (Evanston: Northwestern University Press, 1993).
3. We will not enter into a discussion of Descartes' treatise on light in this chapter (see Descartes [1637] 1965). It must also be noted that the topic of depiction of 3-D objects on a 2-D surface has elicited a great deal of debate in the philosophy of art. See, for example, Gombrich (1977), Goodman (1968), Wollheim (1987) and Hyman (2006).

4. For Descartes, even though bodies are extended, and therefore always in space, the experience we have of space can never lead to true knowledge of it. This is in keeping with Descartes' general privileging of reason over experience, in, for example, the *Meditations on First Philosophy* (Descartes [1641] 1996).
5. See, for example, http://www.tate.org.uk/art/artworks/cezanne-still-life-with-water-jug-n04725. Last accessed 4 November, 2013.
6. For a more detailed discussion of the formative power of lines, see Hoel (2012).
7. For a discussion of the ontological power of technologies and other forms of expression, see Hoel (2011), Hoel and van der Tuin (2013), Carusi (2012) and Carusi and Hoel (2014).
8. See, for example, http://www.metmuseum.org/Collections/search-the-collections/110000310; http://www.metmuseum.org/Collections/search-the-collections/110000311; http://www.courtauld.ac.uk/gallery/collections/paintings/imppostimp/cezanne_montagne.shtml; and http://www.nationalgalleries.org/collection/artists-a-z/C/2913/artist_name/Paul Cézanne/record_id/2488. All websites last accessed 4 November, 2013.
9. Merleau-Ponty makes this point by approvingly paraphrasing André Malraux (1993, 91 ILVS).
10. See Latour (1987) for a classic account of black-boxing in scientific laboratories.
11. We deal with this in more detail in Carusi and Hoel (2014) and in our ongoing work *The Measuring Body*, which develops an account of scientific vision drawing on Merleau-Ponty's later work.

REFERENCES

Alač, Morana. 2011. *Handling Digital Brains: A Laboratory Study of Multimodal Semiotic Interaction in the Age of Computers*. Cambridge, MA: MIT Press.
Barthes, Roland. (1980) 1981. *Camera Lucida: Reflections on Photography*. Translated by Richard Howard. New York: Hill.
Beaulieu, Anne. 2000. *The Space inside the Skull: Digital Representations, Brain Mapping and Cognitive Neuroscience in the Decade of the Brain*. PhD diss., University of Amsterdam.
Beaulieu, Anne. 2001. "Voxels in the Brain: Neuroscience, Informatics and Changing Notions of Objectivity." *Social Studies of Science* 31 (5): 635–680.
Bolter, Jay David, and Richard Grusin. 1999. *Remediation: Understanding New Media*. Cambridge, MA: MIT Press.
Bryson, Norman. 1983. *Vision and Painting: The Logic of the Gaze*. Houndsmills: Macmillan.
Carusi, Annamaria. 2012. "Making the Visual Visible in Philosophy of Science." *Spontaneous Generations* 6 (1): 106–114. Accessed July 15, 2013.
Carusi, Annamaria, and Aud Sissel Hoel. 2014. "Toward a New Ontology of Scientific Vision." In *Representation in Science Revisited*, edited by Catelijne Coopmans, Janet Vertesi, Michael Lynch and Steve Woolgar, 201–221. Cambridge, MA: MIT Press.
Crary, Jonathan. 1990. *Techniques of the Observer: On Vision and Modernity in the Nineteenth Century*. Cambridge, MA: MIT Press.
Damisch, Hubert. 1995. *The Origin of Perspective*. Cambridge, MA: MIT Press.
de Rijcke, Sarah, and Anne Beaulieu. 2014. "Networked Neuroscience: Brain Scans and Visual Knowing at the Intersection of Atlases and Databases." In *Representation in Science Revisited*, edited by Catelijne Coopmans, Janet Vertesi, Michael Lynch and Steve Woolgar, 131–152. Cambridge, MA: MIT Press.

Descartes, René. (1637) 1965. *Discourse on Method, Optics, Geometry, and Meteorology*. Translated by Paul J. Olscamp. Indianapolis: Bobbs-Merrill.

Descartes, René. (1641) 1996. *Meditations on First Philosophy*. Edited by John Cottingham. Cambridge: Cambridge University Press.

Dumit, Joseph. 2004. *Picturing Personhood: Brain Scans and Biomedical Identity*. Princeton, NJ: Princeton University Press.

Enchev, Yavor. 2009. "Neuronavigation: Genealogy, Reality and Prospects." *Neurosurgical Focus* 27 (3): E11.

Gombrich, Ernst. 1977. *Art and Illusion*. Oxford: Phaidon.

Goodman, Nelson. 1968. *Languages of Art*. Indianapolis: Hackett.

Goodwin, Charles. 1994. "Professional Vision." *American Anthropologist* 96 (3): 606–633.

Grasseni, Cristina, ed. 2009. *Skilled Visions: Between Apprenticeship and Standards*. New York: Berghahn Books.

Haraway, Donna. 1988. "Situated Knowledges: The Science Question in Feminism and the Privilege of Partial Perspective." *Feminist Studies* 14 (3): 575–599.

Hoel, Aud Sissel. 2011. "Thinking 'Difference' Differently: Cassirer versus Derrida on Symbolic Mediation." *Synthese* 179 (1): 75–91.

Hoel, Aud Sissel. 2012. "Lines of Sight: Peirce on Diagrammatic Abstraction." In *Das bildnerische Denken von Charles S. Peirce*, edited by Franz Engels, Moritz Queisner and Tullio Viola, 253–271. Berlin: Akademie Verlag.

Hoel, Aud Sissel, and Iris van der Tuin. 2013. "The Ontological Force of Technicity: Reading Cassirer and Simondon." *Philosophy and Technology* 26 (2): 187–202.

Hyman, John. 2006. *The Objective Eye: Color, Form and Reality in the Theory of Art*. Chicago: University of Chicago.

Jay, Martin. 1994. *Downcast Eyes: The Denigration of Vision in Twentieth-Century French Thought*. Berkeley: University of California Press.

Johnson, Galen A., ed. 1993. *The Merleau-Ponty Aesthetics Reader: Philosophy and Painting*. Translated by Michael B. Smith. Evanston: Northwestern University Press.

Joyce, Kelly A. 2008. *Magnetic Appeal: MRI and the Myth of Transparency*. Ithaca: Cornell University Press.

Lanham, Richard. 1993. *The Electronic Word: Democracy, Technology, and the Arts*. Chicago: University of Chicago Press.

Latour, Bruno. 1987. *Science in Action: How to Follow Scientists and Engineers through Society*. Cambridge, MA: Harvard University Press.

Lindseth, Frank, Thomas Langø, Tormod Selbekk, Rune Hansen, Ingerid Reinertsen, Christian Askeland, Ole Solheim, Geirmund Unsgård, Ronald Mårvik and Toril A. Nagelhus Hernes. 2013. "Ultrasound-Based Guidance and Therapy." In *Advancements and Breakthroughs in Ultrasound Imaging*, edited by Gunti P. P. Gunarathne, 27–82. ISBN: 978–953–51–1159–7. InTech. doi: 10.5772/55884. Accessed July 19, 2013. http://www.intechopen.com/books/advancements-and-breakthroughs-in-ultrasound-imaging/ultrasound-based-guidance-and-therapy. Last accessed 4 November, 2013.

Lynch, Michael. 1985. "Discipline and the Material Form of Images: An Analysis of Scientific Visibility." *Social Studies of Science* 15 (1): 37–66.

Merleau-Ponty, Maurice. (1945) 1962. *Phenomenology of Perception*. Translated by Colin Smith. London: Routledge and Kegan Paul.

Merleau-Ponty, Maurice. (1945) 1993. "Cézanne's Doubt." In Johnson, *Merleau-Ponty Aesthetics Reader*, 59–75.

Merleau-Ponty, Maurice. (1952) 1993. "Indirect Language and the Voices of Silence." In Johnson, *Merleau-Ponty Aesthetics Reader*, 76–120.

Merleau-Ponty, Maurice. (1961) 1993. "Eye and Mind." In Johnson, *Merleau-Ponty Aesthetics Reader*, 121–149.

Merleau-Ponty, Maurice. (1964) 1968. *The Visible and the Invisible*. Edited by Claude Lefort. Translated by Alphonso Lingis. Evanston: Northwestern University Press.

Nagel, Thomas. 1986. *The View from Nowhere*. Oxford: Oxford University Press.

Panofsky, Erwin. 1997. *Perspective as Symbolic Form*. Translated by Christopher S. Wood. New York: Zone.

Pustilnik, Amanda. 2009. "Violence on the Brain: A Critique of Neuroscience in Criminal Law." *Wake Forest Law Review* 44: 183–248.

Sturken, Marita, and Lisa Cartwright. 2001. *Practices of Looking: An Introduction to Visual Culture*. Chicago: University of Chicago Press.

Thornton, Davi Johnson. 2011. *Brain Culture: Neuroscience and Popular Media*. New Brunswick: Rutgers University Press.

van Fraassen, Bas C. 2008. *Scientific Representation: Paradoxes of Perspective*. Clarendon: Oxford University Press.

Willats, John. 1997. *Art and Representation: New Principles in the Analysis of Pictures*. Princeton, NJ: Princeton University Press.

Wollheim, Richard. 1987. *Painting as an Art*. London: Thames and Hudson.

8 A Four-Dimensional Cinema
Computer Graphics, Higher Dimensions and the Geometrical Imagination

Alma Steingart

"The merit of speculations on the fourth dimension . . . is chiefly that they stimulate the imagination, and free the intellect from the shackles of the actual."—Bertrand Russell ([1904] 1988, 580)

"Perhaps somebody may appear on the scene some day who will devote his life to it, and be able to represent to himself the fourth dimension."—Henri Poincaré ([1905] 1952, 51)

Thomas Banchoff is consumed by the life and work of Edwin Abbott Abbott. A nineteenth-century English schoolmaster, Abbott is most famous for his satirical novella *Flatland: A Romance of Many Dimensions*, which, since it was published in 1884, became known not for its criticism of Victorian society but for its treatment of higher dimensions. The story is narrated by A Square, a resident of Flatland, a two-dimensional world whose inhabitants are circles, polygons and lines. On the eve of the millennium, the Square is visited in his home by A Sphere, an inhabitant of Spaceland, who seeks to introduce him to the "Gospel of Three Dimensions." A Sphere tries in vain to reason with the Square, describing to him the properties of three dimensions. Yet the Square remains unconvinced. Finally, frustrated with his effort, he pulls the Square out of Flatland. Opening his eyes, the Square proclaims, "I looked, and, behold, a new world! There stood before me, visibly incorporate, all that I had before inferred, conjectured, dreamed" (Abbott 1885, 122). The experience leads the Square to extrapolate that not only three, but four and even higher dimensions must exist. Operating on the analogy that four dimensions is as perceptible to us as three dimensions is to A Square, the story trains readers in thinking about four dimensions. Because he was not visited by a resident of higher dimensions, Thomas Banchoff has had to content himself with building models of them using computer graphics. This chapter describes his project to manifest four-dimensional objects on two-dimensional computer screens, asking what it reveals about geometrical thinking.

As a boy growing up in Trenton, New Jersey, Banchoff first stumbled upon the idea of higher dimensions while reading a 1943 Marvel comic, "Captain Marvel Visits the World of Your Tomorrow." In the story, boy reporter Billy Batson visits the laboratory of Mr. Kiddin, where he is informed that Kiddin's robot assistants are working on the "seventh, eighth, and ninth dimensions." Upon hearing that, Batson wonders what "happened to the fourth, fifth and sixth dimensions." The question has reverberated for Banchoff ever since. While at Trenton Catholic Boys High School, the fourth dimension became the locus of his geometrical and theological investigations, as he grew increasingly convinced that God is a manifestation of the fourth dimension in our three-dimensional world. Banchoff remained interested in theology as an undergraduate at the University of Notre Dame, where he wrote a paper for one of his courses on the fourth dimension and the Trinity. Nonetheless, as he pursued his mathematical studies, he began approaching the fourth dimension primarily as a geometrical problem (Albers, Alexanderson and Davis, 2011).[1] Throughout his graduate studies at the University of California, Berkeley, while working under the supervision of Shiing-Shen Chern, one of the most famous differential geometers of the twentieth century, Banchoff maintained his interest in higher-dimensional geometry. Yet it was not until he arrived at Brown University in 1967, when he was introduced simultaneously to Charles Strauss and computer graphics, that higher-dimensional geometry became the locus of his professional research.

At the time, Charles Strauss was completing a dissertation in computer graphics under the supervision of Andries van Dam in the Applied Mathematics Group.[2] For his thesis, Strauss developed a program that designed systems of pipes. According to Banchoff, when they met Strauss had "marvelous new techniques," but no interesting problems, while Banchoff had "great problems" and was in need of the capabilities Strauss's program offered (Banchoff 1985, 6). Banchoff's "great problems" concerned how to visualize four-dimensional surfaces and objects via their projections into three-space.[3] The process was analogous to taking a three-dimensional object like a cube and finding out what its shadow on a two-dimensional plane would look like when viewed from different distances and angles, as Abbott did in *Flatland*. Building upon Strauss's "marvelous new techniques," the two began producing some of the very earliest computer graphics films, taking mathematical objects and theories as their subject. Their collaboration lasted for over a decade, during which they produced dozens of films.

The introduction of computer graphics animation to mathematics denotes a novel way of engaging with mathematical theories. Films are, fundamentally, a medium for representing not simply objects but also movement. "Cinema," writes Deleuze, "does not give us an image to which movement is added, it immediately gives us a movement-image. It does give us a section, but a section which is mobile, not an immobile section + abstract movement" (Deleuze 1986, 2–3).[4] By transforming abstract and static mathematical objects into directly observable *events*— what Banchoff terms "phenomena" and Deleuze might term "move-ment-images"—computer graphics reconfigures mathematicians' visual culture.[5] In so doing, it also transforms mathematical practice. Further, computer graphics, I claim, does more than simply represent mathematical phenomena. Instead, I interpret computer animation as a powerful tool with which mathematicians train their *geometrical imaginations*. This use of graphics as a training instrument suggests that a mathematical imagination is not innate, as the word "intuition" might convey, but rather needs to be learned, practiced and cultivated.

Over the last three decades, the use of computer graphics has become omnipresent in both mathematical education and research.[6] However, in the late 1960s and early 1970s, not only was the field of computer graphics still in its infancy but also its application to mathematical problems had only begun. By tracking these early works by Banchoff and Strauss, this chapter discerns some of the implications of using computer graphics to study mathematical problems. Because their work was one of the earliest attempts to apply computer graphics to mathematics, it is a perfect place from which to examine how questions of intuition, practice and persona were reconfigured in light of this new technology.

Science studies scholars have described the myriad ways in which scientific images represent reality: as indexes, symbols, ideal types, inscriptions, simulations, simplifications, shorthands or diagrams of objects, phenomena and theories of the natural world (Cambrosio, Jacobi and Keating 1993; Daston and Galison 1992; Dumit 2004; Kaiser 2005; Lynch and Woolgar 1990). Lorraine Daston and Elizabeth Lunbeck recently have drawn attention to the pervasiveness yet heterogeneity of observation as a scientific practice. A history of observations, they write, demonstrates, among other things, "how the senses have been schooled and extended" and "how the private experiences of individuals have been made collective and turned into evidence" (Daston and Lunbeck 2011, 2). That is, it is possible, even necessary, to read histories of scientific representation in light of allied literature about how scientists learn to see, apprehend and interpret data, a body of literature indebted to the practice turn and its emphasis on how scientists' bodies are trained and articulated (on the training of scientific bodies, see Latour 2004; Myers 2008; Rasmussen 1999). Following Banchoff's work, I claim that the application of computer graphics to mathematics is one place from which to examine the intersection of histories of scientific representation with histories of scientific observation. As I make clear, the graphic computer functions in mathematics as both a representational and an observational tool; graphically simulating mathematical objects on computer screens renders them perceptible to geometers.[7] In the process, observations become not only a way of rendering specific phenomena but also a method for cultivating mathematicians' imaginations.

I first describe the early films Banchoff and Strauss produced, arguing that by imbuing the mathematical objects they were investigating with time and movement, the films enabled a completely novel way of engaging with mathematical objects, one that was qualitatively different from static representations. Specifically, animating mathematical objects imbued them with "liveliness," and enabled mathematicians to approach them as manifest things in the world.[8] The computer was transformed from a representational tool, one that could represent complicated images with great accuracy and speed, to an observational one with which mathematicians observed complex mathematical surfaces. I then suggest that the application of computer graphics to mathematics alters the work of mathematicians by exercising their observational faculties. Thus, animating mathematics on computer screens both constructed abstract mathematical objects as entities that could move through time and space and enabled mathematicians to approach them not only cognitively but also sensorially.

Next I show that more than simply displaying mathematical theories and objects, computer graphics expands what can be visualized and perceived in mathematics. Computer graphics thereby functions as a tool for training mathematicians' geometrical imaginations. By describing this capacity as "imagination," I mean to denote the way mathematicians learn to think about mathematical concepts as concrete things in the world, objects that can be approached not just analytically but also sensuously. By way of conclusion, I return to Abbott and his British contemporaries, who insisted that a geometrical object is legitimate only if one can conceive of it visually rather than in its formal definition. Hence they rejected higher-dimensional geometry. Reading computer graphics within the legacy of mathematical thinking about higher dimensions, I suggest that computer graphics extends the horizon of the visually conceivable.

THE TORUS AND THE SQUARE:
LIVELY PROJECTIONS ON THE GRAPHIC COMPUTER

After graduating from the University of California, Berkeley in 1964, Banchoff spent two years at Harvard as a Benjamin Peirce Fellow. He gave up his third year in order to spend a year in Amsterdam as a postdoctoral research assistant for differential geometer Nicolaas Kuiper. While in Amsterdam, Banchoff received an unsolicited letter from Brown University offering him an assistant professorship. He accepted. The move to Brown, where he would meet Charles Strauss, changed the course of Banchoff's career. As already described, Banchoff had higher-dimensional questions in need of a technical fix, and Strauss had a technology in need of a problem. It was, in short, an ideal collaboration.

The two mathematicians set to work developing analytic descriptions of the surfaces they were interested in visualizing. They would then instruct the computer to calculate and produce an image of a given projection. Building upon the computer's data processing power, Banchoff and Strauss could create more accurate images (i.e., the coordinates of the figures were computed analytically). However, this was not what was novel about their approach. What Banchoff and Strauss wanted to animate was a depiction of the shadows in three-dimensional space of a four-dimensional object rotating in four-space.

It took Banchoff and Strauss about a year to produce their first film, *The Hypertorus*. The process was laborious. The film depicts a stereographic projection into three-space of a four-dimensional mathematical object known as the flat torus.[9,10] It took the computer about a minute to calculate each given projection. Each frame consisted of about four hundred vertices, and once the image was complete, Banchoff and Strauss had to photograph the screen and instruct the computer to produce the next projection. At a rate of twenty-four frames per second, it took nearly thirty hours to produce just one minute of film. Banchoff and Strauss then had to send the images they produced to a production company in Boston and wait for the film's arrival. After over forty years, Banchoff has not lost the feeling of awe and excitement he first felt when watching the film for the first time. When reviewing the film in 2011, he explained, "We knew that there was one scene right here that is very exciting because it [the surface] was turning inside out . . . I still get sort of goose bumps from watching this because it was so clear that this is something that no one had ever seen before and we were *seeing* it for the first time" (Thomas Banchoff, pers. comm.). His conversion to computer graphics as a geometrical tool was immediate.

The films Banchoff and Strauss continued to produce during the 1970s were visually quite simple, appearing even primitive from the perspective of contemporary computer graphics. For most of their early work, Banchoff and Strauss used a vector graphics display.[11] This implied that the only way they were able to illustrate figures was with simple lines—no surface areas could be filled in. In Banchoff and Strauss's films, the wire-frame figures always rotate against a uniform black background. Moreover, the wire-frames are almost always white, because adding color could be accomplished only by placing color filters in front of the screen.[12] Nonetheless these visually simple films already suggested the myriad ways in which the application of computer graphics could alter mathematical practice, promoting an experiential understanding of higher-dimensional geometry.

In describing this early work with Strauss, Banchoff recalled, "We were able to produce images which were projections of surfaces in 4-space and to manipulate those projections so as to make *direct visual investigations of phenomena* we previously knew only through formulas and abstract arguments" (Banchoff 1985, 6).[13] The transformation of objects into events allowed Banchoff to ascribe liveliness to the mathematics he was investigating. No longer existing solely in the realms of formal definitions and equations, the objects he examined were brought to "life": it is this process of *becoming* that, four decades later, still raises goose bumps on Banchoff's skin.

When he viewed *The Hypertorus* for the first time, Banchoff was eager to watch a particular sequence in the deformation. The "exciting scene" was a moment in the deformation when the projected surface divides all of three-space into two congruent parts (see Figure 8.1a). Banchoff described the experience of watching it for the first time as "miraculous," adding, "I had never had a thrill like this in my life" (Thomas Banchoff, pers. comm.). He explained, "We knew what was going to happen, but we didn't know what it would look like . . . People made plaster models in the late nineteenth century. They knew this thing stretches out to infinity . . . but this isn't made out of plaster, it really moves. The first time anyone's ever seen it move. *Literally*" (ibid.). His remarks point to the difference Banchoff maintains between computer graphics visualization and older forms of representation, such as illustrations and models. As I show, by enabling movement and providing a new mode of engagement with mathematical objects, the graphic computer functioned as an observational tool more than a representational one.

Banchoff's claim that no one had ever "seen it [the surface] move" suggests that the event is assumed to exist prior to its representation. But the reason no one had ever seen the surface of a projected hypertorus move before was because it never had. The illusion of movement is only engendered by screening static images at twenty-four frames per second. Science studies scholars have analyzed film as a medium of scientific representation, demonstrating how cinema, since its earliest incarnations in the late nineteenth century, has allowed scientists to capture temporal phenomena that cannot be depicted

Figure 8.1   On the left (1a) is a still from Banchoff and Strauss's first film, showing a projection of the flat torus into three-space, which divides the space in two congruent halves. On the right (1b) is a later rendering of the same projection with color and shading. © Thomas F. Banchoff and Davide P. Cervone.

statically. Several scholars have focused on the use of film in the life sciences, demonstrating how films have rendered perceptible biological objects or processes that were otherwise unvisualizable (Cartwright 1995; Landecker 2005, 2006). The early films produced by Banchoff and Strauss were not aimed at slowing down an event that was too slow or too fast; rather their novelty lay in their aim to depict through movement an object that was otherwise understood as static. If Marey's chronophotography studied movement by freezing it in numerous still frames and observing it discontinuously, Banchoff and Strauss's early mathematical films studied objects and theories that had previously been represented only discontinuously by instilling them with movement (Braun 1995). Yet, somewhat surprisingly, the effect is similar to that of early biological films.

In her analysis of early microcinematography, Hannah Landecker argues that in making perceptible cellular processes that were otherwise too slow to be seen, films became experiments in perceiving not simply living things but rather the category of "life." Landecker explains that "after so many years of seeing static drawings, photographs, and fixed and stained slides of cells, scientists voiced surprise and wonder at these scenes of continual movement," noting their "realism and vitality" (Landecker 2005, 919). Banchoff also expresses wonderment, finding the movement of the surfaces he investigated nothing short of "miraculous." In giving mathematical objects movement, the films brought these objects to life, transforming them from abstract formulas into manifest things in the world, entities capable of moving and surprising their observers, who imagined that they preceded their own representation.[14] After years of studying mathematical surfaces through formulas, images and three-dimensional models, Banchoff was astonished by what transpired on the screen: "We *saw* an object from the fourth dimension," he wrote years later, "and we knew that moment that we would never again be satisfied only with static images" (Banchoff 1985, 6).

The cellular film, Landecker continues, "was a new form of narrative as well as a new set of aesthetic forms for both scientists and laymen" (Landecker 2005, 912–913). The films Banchoff and Strauss produced also employed new narrative strategies. The mathematical objects they displayed became "phenomena" to be investigated and observed rather than formulas to be analyzed. But whereas early microcinematography depicted phenomena that had previously been observed in the filmmaker's laboratory under microscopes, the phenomena Banchoff and Strauss's films portrayed were events of their own devising. They were not indexes of any observable event in the natural world. Rather, they only became phenomena at all via their graphic representation and animation. The films, that is, did not represent events in or illustrate theories of the natural world. Banchoff was attracted to computer graphics precisely because it enabled him to escape the confines of nature, not to imitate it. However, investigating mathematical surfaces with the aid of computer graphics did enable Banchoff to approach and investigate higher-dimensional objects as if they were things in the world.

It took Banchoff and Strauss five more years to produce their next two films, *The Hypercube Projections and Slicing* and *Complex Function Graphs*. In order to distribute their films, they founded Banchoff/Strauss Productions, which sold copies of the films and rented them to interested parties. They also screened the films and gave lectures across the country based on their findings. During a 1973–1974 sabbatical at the University of California, Berkeley, Banchoff presented preliminary versions of both films in conjunction with two talks he gave, which piqued the interest of his dissertation advisor, S. S. Chern. After he screened it again two years later at a symposium in Chern's honor, the latter suggested that Banchoff present his films at the upcoming International Congress of Mathematicians in Helsinki.[15] In 1978, before an audience of four hundred mathematicians, Banchoff described the work he and Strauss had done for over a decade (Banchoff 2008). The talk aimed to raise interest among mathematicians in

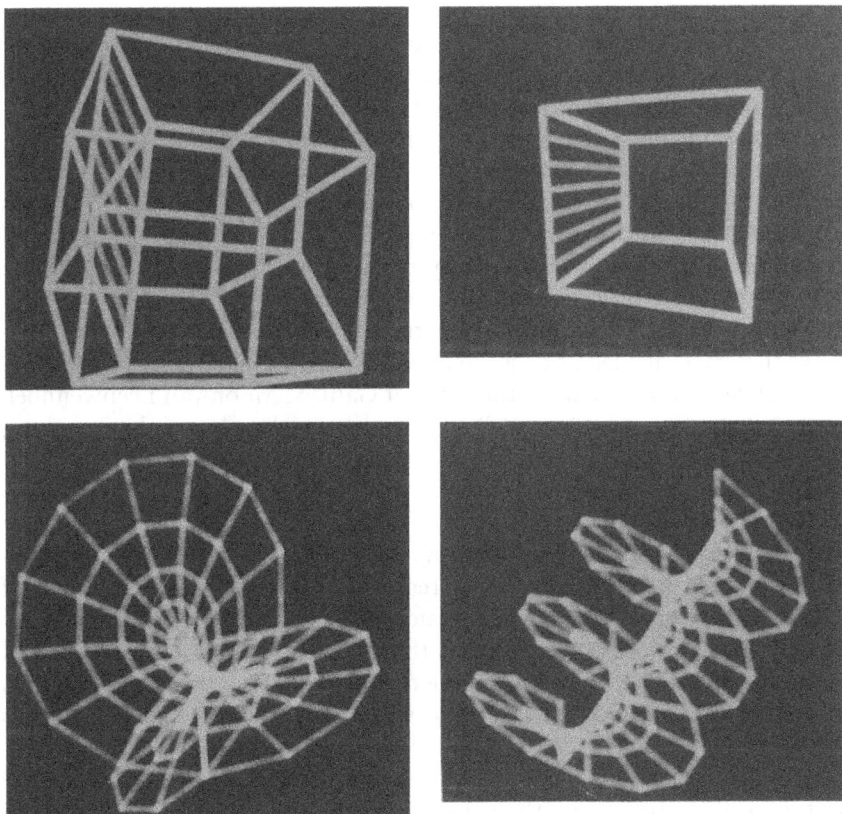

Figure 8.2   On the top (2a & 2b) are two images from *The Hypercube Projections and Slicing*. Below (2c & 2d) are two images from *Complex Functions Graph*. All four images are Polaroid photographs that Banchoff and Strauss took of the computer screen. © Thomas F. Banchoff and Charles Strauss.

the application of computer graphics to both teaching and research. Banchoff presented examples from a number of the films he produced with Strauss, among which was *Hypercube* (see Figure 8.2).[16]

This film is probably the best known and most widely distributed film the two ever produced. It depicts a simple mathematical object, a four-dimensional cube, yet it demonstrates the complexity involved in visually examining a four-dimensional object. In the same way that a cube is an extension of a square into the third dimension, a four-cube is an extension of the cube into the fourth dimension.[17] The film portrays various projections of the four-cube as it rotates in four-space from both a central and an orthogonal projection.[18] Moreover, the film depicts a series of "slices" of the four-cube by a hyperplane, analogous to the way a three-dimensional cube might be sliced by a succession of parallel planes (where the planes can cut perpendicularly into one of the cube's surfaces or at a given angle). Many of the early films the two produced dealt with what Banchoff described as classical problems in complex analysis and differential geometry. For the most part, these were problems mathematicians were familiar with for many years, and could easily be expressed analytically. However, such problems had until then resisted mathematicians' attempts at visualization specifically because they were four-dimensional.

Banchoff describes watching a wire-frame model deform on a computer screen as like peering through a window into a room (Banchoff and Strauss 1978). Yet it is his comparison of computer graphics to older observational techniques that reveals most clearly the way he came to conceive of the computer. In the introduction to his book *Beyond the Third Dimension*, Banchoff writes, "The graphic computer is but the latest in a series of inventions that have enabled us to see in previously inaccessible directions" (Banchoff 1996, 10). Referring to the work of Galileo, Anton van Leeuwenhoek and Wilhelm Conrad Röntgen, Banchoff suggests that the graphic computer is no different than the telescope, the microscope or the X-ray machine. If those technologies enabled scientists to see further into the sky, resolve objects too small to be seen by the naked eye, or peer beneath the surface of the human body, the graphic computer, for Banchoff, allows mathematicians to see higher-dimensional realms. Animating representations of mathematical objects not only transforms them into observable phenomena but in the process also changes the computer from a representational to an observational tool. The computer's work calculating the thousands of still wire-frame models, the vertices, the edges and the transformations between them is forgotten or ignored.[19]

It might be tempting to read Banchoff's analogy as a claim about the ontological status of mathematical objects. By drawing comparisons between the computer and the telescope, Banchoff in effect suggests a similar correspondence exists between a distant star and a hypercube. Yet Banchoff's emphasis is not as much on ascribing realism to the objects he studies as it is on the *way* in which he studies them. That is, the hypercube does not correspond to a star, but the graphic computer allows mathematicians to look at a hypercube in the same way as an astronomer would observe a star. While discussing *Hypercube*, Banchoff remarked that if a scientist wanted to study a frog, he would most likely examine it from various points of view before cutting it open to look inside (Thomas Banchoff, pers. comm.). The ease with which Banchoff switches between talking about the hypercube, a four-dimensional object most people have a hard time conceptualizing, let alone seeing, and the familiar laboratory frog is at first surprising. However, the comparison Banchoff draws is not between the two objects but rather between the ways scientists and mathematicians approach them. Banchoff asks, if scientists use observation to gather more information about a frog, why should mathematicians not do the same? For him, the potential of using computer graphics in mathematics derives from its ability to simulate mathematical objects *as if* they were things in the world. It hence enables mathematicians to approach them as visual phenomena.

What this case suggests is that in mathematics, ontology and epistemology are often entangled. I do not mean this as a statement about mathematical philosophy, but rather as an observation about mathematical practice. Banchoff's investigations demonstrate that for working mathematicians, what you know and how you know it are closely related. Mathematical objects can be defined by an abstract definition, but in mathematical practice, mathematicians approach and investigate the objects of their study in myriad ways—algebraically, symbolically, graphically, pictorially, physically and even through animation. The way a mathematician might think about a given mathematical object is inevitably wrapped up in the manner in which it is represented. This case also demonstrates the ontological multiplicity of mathematical objects as entities that are continuously being "enacted," albeit in different forms, through mathematical practice (Law and Mol, 2002; Mol, 2002).[20]

Banchoff's mathematical publications are deeply influenced by his work with computer graphics. In 1982, Banchoff coauthored *Cusps of Gauss Mappings* with Terence Gaffney and Clint McCroy. The book was inspired by a graphical feature of the early films—namely, because the wire-frame surface projections were transparent, it was easy to see the places where the surface folds or sharply bends. As the surface rotates on the screen, these features come in and out of view. Studying these transformations became a topic of the seminar in which mathematicians presented and studied various examples with the help of computer graphics. The use of wire-frames to represent surfaces, Banchoff explained, implied that "the primary features which lend themselves to mathematical exploration are the folds and cusps of the projected image" (Banchoff 1985, 7). It was a specific visual feature of his films that motivated Banchoff's mathematical study. That is, it was keen observation rather than abstract reasoning that drove Banchoff to new questions and areas of research.

By the early 1980s, Strauss's work as a consultant on mortgage and leasing programs became more demanding and he could no longer continue his collaboration with Banchoff. Luckily for Banchoff, by then many of his students were proficient enough in programming to step in as mathematical filmmakers. At the same time, the technology Banchoff and Strauss had been using was changing: raster graphics displays replaced the older vector graphics displays. This change sharply impacted the appearance of geometrical images—surfaces could be represented in a new manner. Instead of wireframe models, they were solids with color, shading and highlights. These transformations in visualization techniques prompted Banchoff and his students to ask new sorts of questions. No longer interested in folds and cusps, Banchoff now began investigating double and triple points (i.e., points at which two or three surfaces intersect). Banchoff explained:

> We were still looking at the same Veronese surface, sometimes with precisely the same views, but we were led to totally different sorts of conjectures when we saw them in new ways. It was quite a surprising experience for a "pure" mathematician to see to what degree the most natural questions were being determined not by abstract criteria intrinsic to the mathematical subject but rather by the very practical matters of what it was that happened to be easiest to see at the time (Banchoff 1985, 8).

Visualization here motivates analytical work, rather than vice versa. At stake, as Banchoff's admission of surprise reveals, is what it means to *do* mathematics.

Analyzing mathematics as a mode of observation rather than cogitation entails a different articulation of what counts as mathematical work. Approaching mathematical objects as things in the (computer) world calls

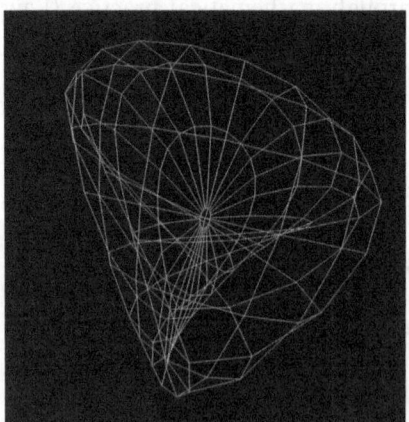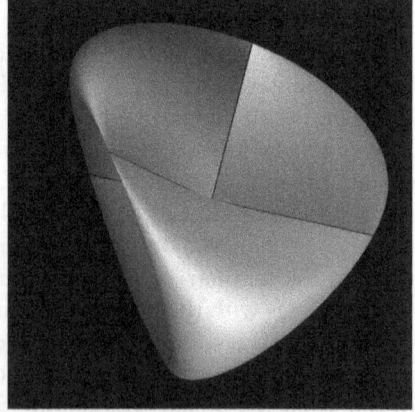

Figure 8.3   Two versions of the Veronese surface. A wire-frame model is on the left (3a) and a more modern rendering is on the right (3b). © Thomas F. Banchoff and Davide P. Cervone.

upon mathematicians' faculties of observation, imagination and perception in addition to their skills in abstract reasoning. Computer graphics animations of mathematical theories and objects not only denote a different mode of apprehension but also signal a new way of producing mathematical knowledge. Visual representations become means of posing new problems and exploring possible solutions.

TRAINING THE GEOMETRICAL IMAGINATION

David Hilbert began the 1932 preface to his book *Geometry and the Imagination* as follows:

> In mathematics, as in any scientific research, we find two tendencies present. On the one hand, the tendency toward abstraction seeks to crystallize the logical relations inherent in the maze of material that is being studied, and to correlate the material in a systematic and orderly manner. On the other hand, the tendency toward intuitive understanding fosters a more immediate grasp of the objects one studies, a live rapport with them, so to speak, which stresses the concrete meaning of their relations (Hilbert and Cohn-Vossen 1952, iii).

In this passage, Hilbert emphasizes both abstraction and "intuitive understanding" as fundamental to mathematical thinking; he continues by emphasizing the importance of a "live rapport" to understanding geometry.

The question of how to wrap one's head around geometrical concepts had already been a sticking point in mathematics at the time of Hilbert's writing. Some geometrical objects and theories lend themselves well to an "intuitive understanding"; others stretch the limits of human comprehension, boggling minds with higher dimensions and unfamiliar geometries. The edge between formalism and intuition is not sharply delineated—indeed, what Banchoff's films demonstrate, in part, is how a previously formal concept can be brought into the realm of human sensibility, in this case via technical prosthetics. Computer graphics, I argue, more than being a representational instrument, can serve as a tool for training mathematicians in how to perceive space, objects and phenomena that otherwise resist three-dimensional spatial understanding. When used in this way, the graphic computer demands convention-based and theory-laden ways of seeing. Banchoff and Strauss's films are devices for training mathematical imaginations, which supplement and augment human cognitive capacity, teaching mathematicians how higher dimensions might be seen, and in what ways.[21]

In their book *The Mathematical Experience*, under the heading "Four-Dimensional Intuition," mathematicians Reuben Hersh and Philip Davis proclaim that an "intuitive grasp of four-dimensional figures is not impossible" (Davis and Hersh 1981, 403). The two go on to describe the work of

Banchoff and Strauss. One of the authors reports that he appreciated the visual aspects of the *Hypercube* upon seeing it for the first time and that he was impressed by Banchoff and Strauss's achievement.[22] However, he remarks that he was left somewhat frustrated, feeling that he did not quite gain an intuitive grasp of the problem. He recounts a visit he made shortly after watching the film to the computing center at Brown University where the film was produced. In front of the computer, Strauss explained to the author how to manipulate the image on the screen via the various handheld controls. He then stood up and let the author take control of the wire-frame model on the screen. He recalls:

> I tried turning the hypercube around, moving it away, bringing it up close, turning it around another way. Suddenly I could *feel* it! The hypercube had leaped into palpable reality, as I learned how to manipulate it, feeling in my fingertips the power to change what I saw and change it back again. The active control at the computer console created a union of kinesthetics and visual thinking which brought the hypercube up to the level of intuitive understanding (Davis and Hersh 1981, 404).

The author's exclamation "I could *feel* it!" and his assertion that the hypercube "leaped into palpable reality" again evince the transformative power of computer graphics, which became, following Hilbert's phrasing, a means of attaining a "live rapport" with the object under investigation. "Intuitive understanding" did not reside in the mathematician, but rather was achieved through active observation and technically mediated manipulation.[23]

This experience at Banchoff and Strauss's lab led Davis and Hersh to conclude that with the advancement of computer graphics, a mathematician could begin with an abstract or algebraic understanding and, by implementing it in a computer program, obtain four-dimensional intuition. The author concludes, "One wonders whether there really ever was a difference in principle between four-dimensional and three-dimensional" (Davis and Hersh 1981, 405). Once a mathematician develops a four-dimensional intuition, the fourth dimension "does not seem that much more imaginary than 'real' things like plane curves and surfaces in space. These are all ideal objects which we are able to grasp both visually (intuitively) and logically" (ibid.). The author echoes Hilbert's two tendencies, but he also points to the moveable boundary between what counts as "real" and "imaginary" in mathematics. According to this interpretation, since mathematicians tend to deal with "ideal objects," their realism has more to do with the way mathematicians approach them than with some quality inherent in them. Put somewhat differently, the only thing that makes a circle more "real" than a four-cube is that a mathematician can *imagine* the former. What the work of Banchoff and Strauss demonstrates is that computer graphics can expand the number of imaginable mathematical objects.

The author notes that it was not until he could actively manipulate the hypercube on the screen that he felt his understanding of the hypercube change. He writes that it was the "union of kinesthetics and visual thinking which brought the hypercube up to the level of intuitive understanding." With the memory and processing power of personal computers today, rotating and "slicing" a four-dimensional object on the screen is computationally trivial. However, when Banchoff and Strauss began collaborating, this was not the case. They were able to observe some transformations on the screen, but others took the computer so long to calculate that the only viable option was to take a photograph of individual frames and then project it as a reel-to-reel film. Further, films were uniquely suitable to disseminating and sharing their work with other mathematicians. The task Banchoff and Strauss faced, therefore, was how to teach the viewer to imagine four-dimensional phenomena by observations alone.

Banchoff acknowledges that before a mathematician can observe higher-dimensional phenomena, he or she must first learn how to *see*. He likens the experience to those of past scientists: "As we watch images moving on the screen of a graphic computer, we are faced with challenges like those of the first scientists to make use of telescopes or microscopes or X-rays. We are seeing things now that have never been visible before" (Banchoff 1996, 11). Banchoff's assertion points to the mediated nature of scientific modes of seeing. In effect, Banchoff here echoes the point made by historians of science who have demonstrated how new tools inculcate new ways of seeing.[24] The introduction of these tools into scientific practice built upon preexisting visual conventions but also extended them, teaching scientists how to understand and interpret new forms of technically mediated imaging (Myers 2008; Rasmussen 1999). The introduction of computer graphics into mathematics is no exception.

The films Banchoff and Strauss produced built upon a long tradition of representing four dimensions through projection into three-space, as well as the use of dimensional analogy. For example, *Hypercube* begins by demonstrating how information about the familiar cube can be obtained through its projection onto a flat, two-dimensional surface. It is then up to the viewer to interpret how that reasoning can be extended to the case of the four-cube. In effect, it is precisely this technique that Edwin Abbott uses in *Flatland*, in which a square living on a two-dimensional surface is introduced to a sphere living in three-space. Yet Banchoff and Strauss sought to build a number of additional conventions into the films they produced. Most notable was their use of standardized progression in the film.

During the 1970s and early 1980s, Banchoff and Strauss produced several films, and while each one visualized a distinct surface, the procedure remained the same in each. For example, each film begins with a three-dimensional wire model rotating uniformly around a vertical axis for a few seconds before the surface begins to deform. This convention creates an illusion of three-dimensionality—otherwise the wire-frame model could be "flattened" onto the screen. Thus, the viewer is led to conceive of the uniform black background on which the figure is drawn not as a flat surface, but rather as a three-dimensional empty space in which the following actions are taking place. From there, the projection progresses "using essentially the same script," by

> introducing a surface in three-space and having it rotate around a vertical axis, then rotating in four-space, stopping that rotation and following it by another familiar three-dimensional rotation and repeating through a sequence of explorations, ending up returning to the original position. The idea is that the audience will be comfortable with the scenario and they will be able to watch a new figure on a sort of "fashion show runway," appreciating the new features without being aware of the background and the sequence of moves by the model (Thomas Banchoff, pers. comm.).

That is, Banchoff and Strauss recognized that the success of the films depended on their ability to cultivate a new way of seeing.[25] A given film was more than a manifestation of a given surface. Just as importantly, it trained the viewer in how to perceive four-dimensional objects in a standardized way.

This last point is especially important when considering that the four-dimensional mathematical objects Banchoff and Strauss sought to represent had been challenging mathematicians for so many years precisely because they *cannot* easily be visualized. After all, the films are three-dimensional representations on two-dimensional screens of four-dimensional objects. This fact suggests that more than being able to see *per se*, the aim of the films was to teach mathematicians how to conceive of, or rather to imagine, higher dimensions.

Computer graphics, more than making new phenomena perceptible, made them imaginable. Here, Banchoff's recollections are instructive:

> My hypothesis was always that after you had a certain amount of experience even with these wire-frame models, your mind really wants them to be continuous, so your mind will really seek a continuous path through this cinematographic space so that you will be watching more than one item. If you are only watching one item and it flattens out then you lose your point of reference, but if you are watching a couple of things simultaneously then you have a choice as to which one will carry you through while the other loses its power . . . you can train yourself to study these things and see things (Thomas Banchoff, pers. comm.).[26]

Watching the films, according to Banchoff, is a regime for training oneself in how to observe. While it may have a tacit component, it is in no way innate. That is, it is experience, practice and habit that a viewer harnesses. The ability simultaneously to watch several features of the surface as it deforms on the screen is not an outcome of analytic calculation or abstract reasoning. It is apprehension and perception that are joined in this practice.

The literature on mathematical intuition is composed of two distinct yet overlapping traditions—a philosophical and a historical one. Philosophers of mathematics are concerned with the question of whether certain mathematical objects or propositions are the result of direct intuition, and to what degree those intuitive understandings lie at the root of all mathematical knowledge. Much of this literature revolves around Kant's notion of the synthetic *a priori* intuition of space, but other mathematical structures such as natural numbers, sets and the continuum have also received philosophical treatment (Giaquinto 2007; Mancosu, Jørgensen and Pedersen 2005; Parsons 1979, 2008; Tieszen 1989). For historians of mathematics, intuition becomes an analytic device through which the historian can discern how changes in epistemological commitments are culturally situated (Gray 2008; Mehrtens 1990, 2004). The appropriation of computer graphics into mathematics, as Banchoff and Strauss's work makes clear, reveals that geometrical intuition can be cultivated and extended. It thus becomes a historical construct in itself. Geometrical intuition, or rather the geometrical imagination of mathematicians at a given time, is built upon a larger complex of visualization technologies and epistemic questions.

CONCLUSION: A RETURN TO THE VICTORIAN

The study of higher-dimensional geometry as an independent field of inquiry emerged in the middle of the nineteenth century with the work of Bernhard Riemann. Until then, most mathematicians had rejected the study of higher-dimensional geometry on the grounds that geometry was framed as the study of physical space. When mathematicians did appeal to *n*-dimensional geometry, they did so only as an analytic extension of specific cases. That is, these extensions were used to solve specific problems, but were not associated with a conceptual, let alone physical, interpretation. Riemann first forwarded a comprehensive treatment of geometry challenging the centrality of three-dimensional Euclidean space. In his 1854 lecture "On the Hypotheses that Lie at the Foundation of Geometry," which was published posthumously three years later, Riemann proposed his now famous metric conception of space, extending the work of Gauss into higher dimensions. Riemann's ideas, however, were not instantly accepted by all mathematicians. This was especially the case in England, where the idea of higher-dimensional geometry had been fiercely debated in the second half of the century, when Abbott was writing his novella. As Joan Richards has shown, for British mathematicians, who for the most part maintained a descriptive approach to geometry, the legitimacy of geometrical ideas did not lie in their formal presentation but rather in their clear conceptualization (Richards 1988).

In 1869, English mathematician James Sylvester addressed the Mathematical and Physical Section of the British Association for the Advancement of Science. The core of the address, which was later published in *Nature* under the title "A Plea for the Mathematician," was a rebuttal to recent remarks made by biologist Julian Huxley regarding the nature of mathematical inquiry. Specifically, Sylvester took issue with Huxley's remark that "mathematics is the study which knows nothing of observation, nothing of experiment, nothing of induction, nothing of causation" (Sylvester 1869, 237). This characterization, according to Sylvester, could not be further from the truth. The article spurred discussion among English mathematicians, not because of Sylvester's refutation of Huxley but rather because of his somewhat peculiar decision to discuss four-dimensional geometry in service of his argument. Sylvester asserted that the notion of generalized space should be a legitimate part of geometry, not just an algebraic formalization.[27] Fifteen years before the publication of *Flatland*, he wrote, "As we can conceive beings (like infinitely attenuated book-worms in an infinitely thin sheet of paper) which possess only the notion of space of two dimensions, so we may imagine beings capable of realizing space of four or a greater number of dimensions" (ibid.). In a footnote, Sylvester admitted that while he was not unable to conceive of "transcendental space," other great mathematicians, such as Gauss, Cayley, Salmon and Clifford, are assured of it. He concluded, "I strive to bring my faculties of mental vision into accordance with theirs" (ibid).

Sylvester's view stood out among his contemporaries, who believed that higher-dimensional spaces did not constitute a legitimate part of geometry. In 1870, Sylvester's views were ridiculed in a *Nature* editorial written by C. M. Ingleby. Ingleby argued that Sylvester "backs up his belief by the names of seven great mathematicians, who are *hypothetically* assumed to have 'an inner assurance of the reality' of space of four dimensions" (Ingleby 1870, 289). "It would be more satisfactory to unbelievers like myself," he added, "if the gifted author of the address were to assure the world that he had an insight into, or clear conception of, this transcendent space" (ibid). In her analysis of this case, Richards remarks that "Ingleby was supported in his belief . . . by virtually all of the English mathematicians who commented on this subject during the ensuing decades" (Richards 1988, 56–57). She adds, "Their focus on epistemological and psychological criteria to judge mathematical validity was universal" (ibid). That is, for English mathematicians in the second half of the nineteenth century, four-dimensional space represented a challenge exactly because it could not be imagined.

This story makes clear the extent to which Abbott's work, with its implicit suggestion that there is no fundamental difference between three- and four-dimensional space, diverged radically from the ideas of his contemporaries. While his was a work of satire rather than pure mathematics, his treatment of dimensions shares much with that of Banchoff, who seeks to make higher dimensions not only conceivable but more importantly, also imaginable. Abbott, or more accurately, the two-dimensional Square who narrates the novella, dedicates *Flatland* to:

The inhabitants of space in general . . .
So the citizens of the Celestial Region
May aspire higher and higher
To the secret of four five or even six dimensions
Thereby contributing
To the enlargement of the imagination (Abbott 1885, i).

Banchoff's films seek a similar goal. In his films, he strives not only to visualize specific mathematical problems but also to promote a certain approach to higher-dimensional geometry, training his viewers in "the enlargement of the imagination." Like the Sphere, he seeks to instill the "gospel of higher dimensions" in his audience. In so doing, he extends the realm of geometry that can be understood not only through formal description but imaginatively as well.

Comparing the study of higher dimensions in Abbott's time with his own, Banchoff writes:

Encounters with phenomena from the fourth and higher dimensions were the fabric of fantasy and occultism. People (other than the spiritualists) did not expect to see manifestations of four-dimensional forms any more than they expected to encounter Lilliputians or Mad Hatters. Today, however, we do have the opportunity not only to observe phenomena in four and higher dimensions, but we can also interact with them. The medium for such interactions is computer graphics (Banchoff 1990, 370).

Banchoff highlights how computer graphics alters mathematicians' engagements with mathematical objects and theories, transforming the fantastical Mad Hatter into a more pedestrian figure easily encountered. Computer graphics did not revolutionize mathematics. A longer history of mathematicians' engagements with higher dimensions shows that the effort to establish a "live rapport" with complex phenomena is not new. Computer graphics did, however, extend the horizon of what sorts of geometrical objects mathematicians could learn to comprehend. The geometrical world has always been populated by infinitely attenuated bookworms, citizens of Celestial Regions, Lilliputians and Mad Hatters (not to mention Mr. Kiddin and his robot assistants). Using computer graphics, mathematicians reclaim the imaginative as the proper domain of geometrical thinking.

ACKNOWLEDGMENTS

The research and writing of this chapter was supported by a grant from the National Science Foundation Dissertation Research Improvement Grant (Award No. 1057311). I would like to thank David Kaiser, Joan Richards, Sophia Roosth and Stefan Helmreich, all of whom carefully read drafts of this chapter and offered insightful comments. Annamaria Carusi, Aud Sissel Hoel and Timothy Webmoor provided critical commentary, pushing me to clarify my thinking on mathematical modeling. Finally, it is a great pleasure to offer my gratitude to Thomas Banchoff, whose work inspired this chapter and who was always generous with his time, source materials and stimulating conversation.

NOTES

1. The biographical material regarding Banchoff's childhood and school years is also based on personal interviews with the author.
2. Van Dam was one of the American pioneers in the field of computer graphics. In 1967, he cofounded the Association of Computer Machinery's SIC-GRAPH, which is the precursor to SIGGRAPH (Special Interest Group on Computer Graphics and Interactive Techniques).
3. To depict a mathematical object or surface in film, Banchoff and Strauss had to decide first what sort of projection techniques to use (often they used more than one in a given film), and then the angle of rotation. Each configuration, by supplying the viewer with an alternative perspective, can reveal different information about the object.
4. Deleuze builds his analysis upon a reading of Bergson's *Matter and Memory* and *Creative Evolution* (Bergson 1911, 1913).
5. For a theoretical account comparing the history of cinema to computer graphics and other new media, see Manovich (2002).
6. The appropriation of computer graphics into mathematical research and education increased over the 1980s and 1990s. In 1991, for example, the National Science Foundation founded the Geometry Center, a research institute dedicated to bringing mathematicians and computer graphics specialists together to solve visualization problems in pure mathematics. The center closed seven years later. However, many of the mathematicians involved in the project continued to advance the use of computer graphics in mathematics. In the last decade, Springer Verlag began publishing a series called Mathematics and Visualization, which covers the "application of visualization techniques in mathematics." (http://www.springer.com/series/4562, accessed January 22, 2013). Finally, the Mathematics Subject classification added the category of "visualization" (00A66, together with mathematics and art) to its latest 2010 edition.
7. Throughout his early writings, Banchoff refers to computer graphics as a "graphic computer," because at the time, most computers did not include graphics displays. Computers that did include graphics capabilities, therefore, were termed (at least by Banchoff) "graphic computers."
8. For accounts of how animation imparts "liveliness" to representations, see Myers (2006) and Helmreich (2004).
9. A stereographic projection is a projection from a sphere onto a plane, imagined as the visual effect of a light source placed at the North Pole, with the South Pole located on the plane. The light rays trace where a given point on the sphere will appear on the plane. This means that the closer a surface is to the South Pole, the less distorted it will appear on the plane. The same process can be used to project four-dimensional objects onto three-space. The light source is located at the North Pole of a hypersphere (a four-dimensional sphere) and rays trace the image onto three-space.

10. The flat torus is an embedding of the familiar torus in a four-dimensional Euclidean space. It can be represented analytically, by the following formula: [u,v]=[cos(u+v),sin(u+v),cos(u-v), sin(u-v)/√2] with 0≤u,v≤2p. For a short animation of a stereographic projection of the flat torus, see http://www.geom. uiuc.edu/~banchoff/script/b3d/hypertorus.html, accessed January 22, 2013.

11. The vector and raster graphics displays represent two different modes of producing an image on a cathode-ray tube (CRT). In a vector display, a straight line, also known as a vector, is defined by its two endpoints. Once these two points are specified, an electron beam moves continuously between them and a line is formed on the screen. In a raster graphics display or a television display, the electron beam traces a regular pattern and the intensity of the beam at a given pixel defines the image being displayed on the computer screen. That is, for any individual image, the electron beam must trace the entire screen. The reason vector display was more common in the 1960s and 1970s was that it required less memory to save the image's instructions in the computer. In a vector display, all the information that is needed in order to represent a line is the two endpoints on the line. In a raster display, even for a simple line, the program must hold information regarding each individual pixel in order to determine whether the electron beam should light up.

12. Banchoff and Strauss did use color in their film *The Hypercube*. Each frame was exposed several times, using a different color filter each time, in order to add color to the film. This, of course, prolonged and complicated the production process.

13. Author's emphasis.

14. In a 1999 article entitled "The Visualization of Mathematics: Towards a Mathematical Exploratorium," mathematician Richard Palais, in advocating the use of visualization techniques, explains that in addition to creating accurate representations, with the help of computer graphics it "becomes straightforward to create rotation and morphing animation that can bring the known mathematical landscape to life" (Palais 1999, 647–648).

15. In 1998, a Video Math Festival was held in conjunction with the International Congress of Mathematicians in Berlin (see http://page.mi.fu-berlin.de/ polthier/Events/VideoMath/index.html, accessed January 22, 2013).

16. Banchoff and Strauss's *Hypercube* was not the first computer graphics movie to animate a four-dimensional cube. In 1967, Michael Noll, working at Bell Laboratories, published a paper in the *Communications of the ACM* describing his production of a computer animation of projections of a four-cube (Noll 1967).

17. Where a cube has four vertices and twelve edges, the four-cube has sixteen vertices and thirty-two edges.

18. For an early computer animation of the hypercube, see http://www.youtube.com/watch?v=iXYXuHVTS_k. For a more recent animation, including explanation, see http://alem3d.obidos.org/en/cubeice/movsl4. Accessed January 22, 2013.

19. To a certain degree, Banchoff's ability to approach the graphic computer as an observational tool arises exactly out of this act of black-boxing, whereby the work that goes into representing the object on the screen recedes into the background once the animation begins—the mediation of technology is thereby obscured. In the early days of their collaboration, this distinction between the acts of programming and observing was enhanced by a clear division of labor between Strauss and Banchoff. Strauss was the only one who had the necessary background in computer graphics, while Banchoff focused more on theoretical questions (Latour 1987).

20. In his investigation of diagrams in Greek mathematical texts, Raviel Netz similarly argues that mathematical texts "operate within a multi-layered ontology, with some objects being 'more real' and others being, relatively, 'more imaginary'" (2009, 32). Whereas Netz builds his analysis on a close examination of diagrams and the use of the Greek verb *nenoēsthō* (translated by him as *to imagine*), the early computer graphics films suggest that this layered ontology is also present as mathematical objects are mediated by various technologies. Further, perhaps at play is not simply layered ontology but also layered epistemology. Understanding a four-dimensional surface algebraically is only one way of knowing it; perceiving it visually or kinesthetically is another.

21. In *Where Mathematics Comes From: How the Embodied Mind Brings Mathematics into Being*, George Lakoff and Rafael E. Núñez (2000) offer a cognitive model of mathematics based on theories in linguistics and cognitive science. They argue that conceptual metaphors are at the basis of human mathematical understanding. Their investigation is motivated by philosophical concerns, as they aim to provide a foundational basis for all mathematical knowledge. I am here concerned with mathematical practice rather than philosophy. That is, the question I ask is not "Where does mathematics come from?" but rather "What do mathematicians do?"

22. While the book was written by both authors, several of its chapters are written in the first-person singular. In the preface to their book, the authors explain that in some cases it "will be obvious" which author has written the section, and that in any case, both authors stand behind the text. In this chapter, I refer to the author in the singular to avoid confusion.

23. In her biography of Nobel laureate Barbara McClintock, Evelyn Fox Keller writes, "McClintock's eye was surpassingly well trained. 'Seeing,' in fact, was at the center of her scientific experience" (Keller 1983, 148). Keller further remarks that for McClintock, the "reciprocity between the cognitive and visual seems always to have been more intimate than it is for most. As if without distinguishing between the two, she knew by seeing, and saw by knowing" (ibid.). For Banchoff, it was early computer graphics that provided him with an opportunity to intimately join together ways of "knowing" and ways of "seeing."

24. In *Objectivity*, Lorraine Daston and Peter Galison write, for example, that "in the case of atlases that present images from new instruments, such as bacteriological atlases of the late nineteenth century and the x-ray atlases of the early twentieth century, everyone in the field addressed by the atlas must begin to learn to 'see' anew" (Daston and Galison 2007). See also Crary (1992).

25. The early films did not have a voiceover to help orient the viewer or explicate the source of the depicted movement. The one exception is *Complex Function Graphs*. After the film was completed, Banchoff screened the film in California at Charles Eames's studio. As the film was rolling, Banchoff described to the audience what was taking place on the screen. When the film finished, Eames asked Banchoff if he had a recorded version of his narration as well. Banchoff replied in the negative, and Eames offered that Banchoff use his studio, which he did. The "script" Banchoff and Strauss hoped to establish, therefore, was doubly important because it is not obvious to the uninitiated viewer how the various deformations relate to one another.

26. By the word "choice," Banchoff does not imply that the decision is a conscious one. On the contrary, he believes that "the selection of which part of an object to watch is something done automatically" (Thomas Banchoff, pers. comm.).

27. Sylvester pointed to the work of Riemann, who he claimed showed "that the basis of our conception of space is purely empirical," and that "other kinds of space might be conceived to exist, subject to laws different from those which govern the actual space in which we are immersed" (Sylvester 1869).

REFERENCES

Abbott, Edwin A. 1885. *Flatland: A Romance of Many Dimensions*. Boston: Roberts Brothers.

Albers, Donald J., Gerald L. Alexanderson, and Philip J. Davis, eds. 2011. *Fascinating Mathematical People: Interviews and Memoirs*. Princeton, NJ: Princeton University Press.

Banchoff, Thomas F. 1985. "Computer Graphics Applications to Geometry: 'Because the Light Is Better Over Here.'" In *The Merging of Disciplines: New Directions in Pure, Applied, and Computational Mathematics*, edited by Richard E. Ewing, Kenneth I. Gross, and Clyde F. Martin, 1–14. New York: Springer-Verlag.

———. 1990. "From *Flatland* to Hypergraphics: Interacting with Higher Dimensions." *Interdisciplinary Science Reviews* 15 (4): 364–372.

———. 1996. *Beyond the Third Dimension: Geometry, Computer Graphics, and Higher Dimensions*. New York: Scientific American Library.

———. 2008. "Computer Graphics in Mathematical Research, from ICM 1978 to ICM 2002: A Personal Reflection." In *Mathematical Software: Proceedings of the First International Congress of Mathematical Software: Beijing, China, 17–19 August 2002*, edited by Arjeh Cohen, Kiao-Shan Gao, and Nobuki Takayama, 180–189. Berlin: World Scientific.

Banchoff, Thomas F., and Charles M. Strauss. 1978. "Real-Time Computer Graphics Analysis of Figures in Four-Space." In *Hypergraphics: Visualizing Complex Relationships in Art, Science, and Technology*, edited by David W. Brisson, 159–167. Boulder: Westview Press.

Bergson, Henri. 1911. *Creative Evolution*. New York: Henry Holt.

———. 1913. *Matter and Memory*. New York: Macmillan.

Braun, Marta. 1995. *Picturing Time: The Work of Etienne-Jules Marey (1830–1904)*. Chicago: University of Chicago Press.

Cambrosio, Alberto, Daniel Jacobi, and Peter Keating. 1993. "Ehrlich's 'Beautiful Pictures' and the Controversial Beginnings of Immunological Imagery." *Isis* 84 (4): 662–699.

Cartwright, Lisa. 1995. *Screening the Body: Tracing Medicine's Visual Culture*. Minneapolis: University of Minnesota Press.

Crary, Jonathan. 1992. *Techniques of the Observer: On Vision and Modernity in the Nineteenth Century*. Cambridge, MA: MIT Press.

Daston, Lorraine, and Peter Galison. 1992. "The Image of Objectivity." *Representations* 40: 81–128.

———. 2007. *Objectivity*. New York: Zone Books.

Daston, Lorraine, and Elizabeth Lunbeck. 2011. "Introduction: Observations Observed." In *Histories of Scientific Observation*, edited by Lorraine Daston and Elizabeth Lunbeck, 1–10. Chicago: University of Chicago Press.

Davis, Philip J., and Reuben Hersh. 1981. *The Mathematical Experience*. Boston: Birkhäuser.

Deleuze, Gilles. 1986. *Cinema 1: The Movement-Image*. Minneapolis: University of Minnesota Press.

Dumit, Joseph. 2004. *Picturing Personhood: Brain Scans and Biomedical Identity*. Princeton, NJ: Princeton University Press.

Giaquinto, Marcus. 2007. *Visual Thinking in Mathematics: An Epistemological Study.* Oxford: Oxford University Press.

Gray, Jeremy. 2008. *Plato's Ghost: The Modernist Transformation of Mathematics.* Princeton, NJ: Princeton University Press.

Helmreich, Stefan. 2004. "The Word for World Is Computer: Simulating Second Natures in Artificial Life." In *Growing Explanations: Historical Perspectives on Recent Science,* edited by Norton M. Wise, 271–300. Durham, NC: Duke University Press.

Hilbert, David, and Stephan Cohn-Vossen. 1952. *Geometry and the Imagination.* Providence: AMS Chelsea.

Ingleby, C. M. 1870. "Transcendent Space." *Nature* 1 (January 18): 289.

Kaiser, David. 2005. *Drawing Theories Apart: The Dispersion of Feynman Diagrams in Postwar Physics.* Chicago: University of Chicago Press.

Keller, Evelyn Fox. 1983. *A Feeling for the Organism: The Life and Work of Barbara McClintock.* New York: W. H. Freeman and Company.

Lakoff, George, and Rafael E. Núñez. 2000. *Where Mathematics Comes From: How Embodied Mind Brings Mathematics into Being.* New York: Basic Books.

Landecker, Hannah. 2005. "Cellular Features: Microcinematography and Film Theory." *Critical Inquiry* 31 (4): 903–937.

———. 2006. "Microcinematography and the History of Science and Film." *Isis* 97 (1): 121–132.

Latour, Bruno. 1987. *Science in Action: How to Follow Scientists and Engineers Through Society.* Cambridge, MA: Harvard University Press.

———. 2004. "How to Talk about the Body? The Normative Dimension of Science Studies." *Body and Society* 10 (2–3): 205–229.

Law, John, and Annemarie Mol. 2002. *Complexities: Social Studies of Knowledge Practices.* Durham, NC: Duke University Press.

Lynch, Michael, and Steve Woolgar, eds. 1990. *Representation in Scientific Practice.* Cambridge, MA: MIT Press.

Mancosu, Paolo, Klaus Frovin Jørgensen and Stig Andur Pedersen. 2005. *Visualization, Explanation and Reasoning Styles in Mathematics.* Norwell: Springer.

Manovich, Lev. 2002. *The Language of New Media.* Cambridge, MA: MIT Press.

Mehrtens, Herbert. 1990. *Moderne Sprache Mathematik: Eine Geschichte des Streits um die Grundlagen der Disziplin und des Subjekts formaler Systeme.* Frankfurt am Main: Suhrkamp.

———. 2004. "Mathematical Models." In *Models: The Third Dimension of Science,* edited by Nick Hopwood and Soraya de Chadarevian, 276–304. Stanford, CA: Stanford University Press.

Mol, Annemarie. 2002. *The Body Multiple: Ontology in Medical Practice.* Durham, NC: Duke University Press.

Myers, Natasha. 2006. "Animating Mechanism: Animations and the Propagation of Affect in the Lively Arts of Protein Modelling." *Science Studies* 19 (2): 6–30.

———. 2008. "Molecular Embodiments and the Body-Work of Modeling in Protein Crystallography." *Social Studies of Science* 38: 163–199.

Netz, Raviel. 2009. "Imagination and Layered Ontology in Greek Mathematics." *Configurations* 17 (1): 19–50.

Noll, Michael A. 1967. "A Computer Technique for Displaying n-Dimensional Hyperobjects." *Communications of the ACM* 10 (8): 469–473.

Palais, Richard. 1999. "The Visualization of Mathematics: Towards a Mathematical Exploratorium." *Notices of the American Mathematical Society* 46: 647–658.

Parsons, Charles. 1979. "Mathematical Intuition." *Proceedings of the Aristotelian Society* 80: 145–168.

————. 2008. *Mathematical Thought and Its Objects*. Cambridge: Cambridge University Press.

Poincaré, Henri. (1905) 1952. *Science and Hypothesis*. New York: Dover.

Rasmussen, Nicolas. 1999. *Picture Control: The Electron Microscope and the Transformation of Biology in America, 1940–1960*. Stanford, CA: Stanford University Press.

Richards, Joan L. 1988. *Mathematical Visions: The Pursuit of Geometry in Victorian England*. Boston: Academic Press.

Russell, Bertrand. (1904) 1988. "Review of *The Fourth Dimension* by Charles Howard Hinton." In *The Collected Papers of Bertrand Russell*, Vol. 4, *Foundations of Logic, 1903–05*, edited by Alasdair Urquhart and Albert C. Lewis, 578–580. New York: Routledge.

Sylvester, John J. 1869. "A Plea for the Mathematician." *Nature* 1: 237.

Tieszen, Richard L. 1989. *Mathematical Intuition: Phenomenology and Mathematical Knowledge*. Norwell, MA: Springer.

Part II

Doing Visual Work
in Science Studies

9 Visual STS

Peter Galison

INTRODUCTION: VISUAL SCIENCE AND TECHNOLOGY STUDIES (VSTS)

Behind the most significant accomplishments of the last thirty years of science and technology studies (STS)—behind laboratory studies and actor network theory, at the center of our ventures into scientific intellectual property, authorship, historical epistemology, media studies, book history, discourse analysis, participant-observation and the philosophy of experimentation—in back of all this is a turn toward *locality*.[1] Divided as the various approaches may be about methods and priorities, our complex of disciplines is no longer satisfied with global claims about universal norms and transhistorical markers of demarcation. Localization was key in Steven Shapin and Simon Schaffer's (1985) examination of Boyle's Pall Mall laboratory in its social, class and political location; localization is the sine qua non of Ian Hacking's (1983) insistence on practical intervention—specific procedures at the Stanford Linear Accelerator Center, not Mertonian norms full stop. "*So far as I'm concerned,*" Hacking wrote, "*if you can spray [positrons onto a niobium ball] they are real.*" Harry Collins (1981) aimed for a local account of repetition when he challenged textbook truisms by showing how crucial person-to-person contact was for the replication of a laser. In different ways and with different tools, the STS field-cluster has utterly rejected the intellectual historical conceit that ideas leapt from great book to great book, or arose from a miasmic Zeitgeist.

Tools for excavating the local have come from across the disciplinary map. Ethnographers and sociologists, historians and philosophers, gender theorists and media historians—all have wanted to know about the shifting, productive nature of scientific practice in *particular* times and places.[2] Work increasingly crosscuts these genres—historically inflected ethnography, for example, takes present-day laboratories-in-the-world, expands sources to new forms of the (digital) archive, observation-participation, locates the present within a historical trajectory.

But here and there, something new is emerging from within the local: attention to the visual—as source, evidence and form of reasoning. We are by now familiar with the idea that diagrams, charts, maps and photographs can serve as a fundamental part of scientists' argumentation. What about science and technology studies itself? Can the visual function there too as a form of research, not just popularization or illustration?

Can documentary film, for example, be scholarship—can it be something more than another means of advancing the public understanding of science, more than another box of raw material? Is the analytical power of propositional concepts the very essence of research, and do words alone leave nothing out, other than superficial reportage or arbitrary art? Can there be a kind of knowledge, an epistemological contribution from film that supplements and enriches our understanding of science-in-practice? For all too long, many in the scholarly community have held film (and now the more capacious category of digital-visual material) at arm's length—the sneaking text scholar's suspicion lingers that the visual might be irredeemably incapable of explanation.

In 1960, Jean Rouch, one of the great ethnographic filmmakers, teamed up with sociologist Edgar Morin to produce *Chronicle of a Summer*, a defining contribution to film history in technique (it introduced dual-system [portable] sound, making sync sound possible in the field); in genre (it launched cinema verité, following a group of young French workers and students in the summer of 1960); and in structure (it proffered reflexivity by including filmed responses of participants in the film as they watched a first version). Marxist sociologist Lucien Goldmann responded acerbically. Hegel, he explained, had long ago shown "the truth is the whole"—and these filmic glimpses into particularity were nothing if not partial. Goldmann judged "Chronicle's" conversations, recorded in the streets, factories and homes of Paris, to be bits unintegrated into "global structures," and so "extremely poor in relation to the more complex structure of reality."[3] We see in the film: a young man, lost in the shifting economy of 1960 awakening in his bedroom, his mother bringing him breakfast; workers laboring on the production line; university students trying to find their way; a young woman, a concentration camp survivor, wandering through Paris and remembering her father, whom the Nazis had murdered. Toward the end of the film Rouch and Morin portray these and other participants watching the film in which they have just appeared and responding to it—commenting on their own and others' actions and statements, pushing, tentatively, on the difficult questions of authenticity and representation.

Unmoved, Goldmann judged the film fatally particular, *too local* since such sequences were untied to a sociological "global context." For Goldmann, it came to this: "The film maker, who does not have the possibility of filming concepts . . . can only seize [the larger context] through its reproduction at the level of individual beings and concrete situations which are the only things directly available to him: which is to say, on the level of fiction."[4] For Goldmann, nonarbitrary reality was accessible *only* by mediating ever more *abstract concepts* and these were, irreducibly, unavoidably textual. Citing Marx and Hegel when it came to human realities, concepts were essential: "The truth is never immediate."[5] Goldmann's cutting final judgment: Inevitably local, cinema verité was forever condemned, in virtue of its particularity, to the status of mere fiction.

Since locality, materiality and particularity are indeed the hallmarks that have become, for our present moment of scholarship, central to the very fabric of science and technology studies, we can ask the next question: What *kind* of locality does film afford? I will argue that visual locality extends and complements textual locality—and as such opens the possibility of a generative, visual scholarship that puts on offer precisely what Goldmann disliked, a dense, unexpected immediacy that offers new forms of context. In the following sections, I shall turn to a spectrum of works in STS and visual anthropology that might pencil in the opening contours of this new register of STS—both uses by scientists of images (which I will call a first-order VSTS) and uses of film and media studies to explore scientific practice (which I will refer to as second order). In the final section at the request of the editors, I want to turn to my own trajectory into film and digital media as a way of complementing written analysis. Perhaps thinking through some of what has worked—and some of what has not—might be of use to others.

VISUAL SCHOLARSHIP IN THE WORLD OF WORDS

Against Goldmann, the interpretive social sciences and allied branches of philosophical, literary and art historical studies have embraced the local. Impatient with a rigid commitment to behavior and reason as algorithmic rule-following and with a fierce patrolling of scientific borders, historians, sociologists and philosophers turned to more instantiated knowledge. Back in 1962, Thomas S. Kuhn, reaching for the postanalytic Wittgenstein, pointed to the way scientists followed exemplars (e.g., Newton's calculation of planetary orbits) rather than simple free-floating propositions (e.g., force is proportional the product of masses and inversely to the square of the distance).[6] Piecewise transported to other contexts, eighteenth- and nineteenth-century astronomers could calculate other phenomena, and the ability to apply examples in this way came to define (for Kuhn) a scientific community. Exemplary problem solutions picked out the right kinds of laws and entities with which to proceed. Clifford Geertz agreed—only the densely *instantiated* aspect of culture was of interest as he took up philosopher Gilbert Ryle's (1968) notion of a thick description. The local, social and conventional context gave meaning to an event where the event alone would be undefined or ambiguous. Only this denser context could distinguish an eye twitch from a conspiratorial wink. In Geertz's hands, thick description blossomed into a way of doing anthropology: A Balinese cockfight (1973, 412–454), with all its complexities of convention, in-group solidarity and regulated behavior, captured a local culture in its density. No mere set of covering rules and universal structures governing kinship or wealth transfer could capture this overlay of enacted, symbolized and agreed-upon meanings.

Film offers an approach to the material and social world of science that complements the work of written material foregrounding the local. Film and other visual media subvert text in productive ways. And yet within science studies, film has played only a glancing, secondary role, serving either as source material or as popularization. This is so even at the level of exhibitions: Whereas an art museum presenting an exhibit on Rembrandt's school addresses a mixed audience of engaged viewers, a science museum addresses schoolchildren. Films about science, even where they do not just address the yellow school bus, do see themselves as a way of increasing public understanding of results.

Why? Among historians, perhaps it is because history so often prides itself—in a way not too different from Goldmann—on explaining scientifically its object of study. Writing, so it is assumed, excels at characterizing an episode, partitioning it into periods and explaining it by an ordering externality: economics, psychology, bureaucracy, political allegiance. In the recent study of science, there is an additional appeal: Writing can contrast whole cultures, splintering them into distinct, complex islands in time or in space, each with its own internally coherent, local system of symbols, values and meanings, each with its cosmogenesis, propagation and kinship structures. Anthropologists have excelled at such holistic inquiry, from Franz Boas's anthropological relativism through Geertz's thick descriptions on Bali (1973, 1983) on to the present. There was a moral-political thrust behind this enterprise: a sense that cultures differed in many ways, even radically, but that in the end there was no hierarchy of culture—no meaningful way of *ranking* Baffin Island relative to Berlin.

Correspondingly, we are by now long familiar with the great switching of *scientific* cultures from Thomas Kuhn (classical to relativistic physics) or Paul Feyerabend (scholastic to Galilean physics). Similarly disjunctive alteration can be found in the epistemic ruptures in the neo-Kuhnian STS work of, for example, Pickering (1984) (S-matrix to field theory), and many others in the 1980s and 1990s. In a Kuhnian mode, we can say, "These two (scientific) cultures are radically disjunct, they differ ontologically, epistemologically, nomologically, precisely as the Boasian analyst would set two (anthropological) cultures apart because they differ in their accounts of origin, social order and reproduction."

Helpful as these radical contrasts can be, they can also override the material and affective phenomenology of practice. When visual anthropologist David MacDougall shows, through extended shots, a Turkana bridewealth negotiation in *Wedding Camels* (1980) we see much more than the results of the camel trading. With time and attentiveness, we see the exchange of looks within families and between them. In the regards and emotions of life in process, in persistent, recognizable bodily gestures, faces, tones of voice, the idea of absolute otherness is, to a certain extent, undermined. MacDougall means this to be so—he precisely wants us to encounter a commonality not only between the groups in discussion but also between them and us (MacDougall 1998). Incipient visual STS does, and could expand on, precisely this bridging effect. Think of (or better look at) the remarkable real and time-lapse images produced by William Newman on a website as he and Lawrence Principe photographed and filmed the making of alchemical icons, like the "Tree of Diana"—a surprising silver dendrite structure that grows from a globular configuration of mercury and silver. Suddenly, what has seemed for generations to reside in the purely metaphorical, lyrical or fantastical world of the alchemist comes closer to us than the distancing of print description could afford. Here is a visually enhanced website that makes a difference.[7]

Novel digital techniques now offer intriguing ways to cut across both text and images. One project, a multiauthored locative investigation into the Zenon Corporation (1945–1960) offers an example. Zenon had the task in those years of developing a classified system of distributed computation, one that is now thought to be a first, large-scale attempt to build a computational network. Half a century ago, company officials stored documents in a basement safe beneath the famous Sullivan building in Chicago. Recently rediscovered, the trove is now, appropriately enough, being examined by a team of architects, historians and other scholars with the goal of producing a tablet-useable augmented-reality display, that will allow proximate and distant viewers to make use of the building's internal sites as well as engaging with the computer historical documents, correspondence and other metadata (Burdick et al. 2012). With layers of inquiry possible, there will be no single path through the material. This is the kind of digital project (I actually don't like the term digital humanities, which seems too restrictive) that immediately suggests a myriad of possible analogue inquiries in laboratory history (for example).

Much older than such relatively new nonlinear, layered projects is the field of visual anthropology. Over almost a century, ethnographic filmmakers have carved out a domain with its own journals, meetings and of course films (see, e.g., Barbash and Taylor [1997]). Indeed, almost a hundred years ago, anthropologists, like Robert Flaherty, began using film to approach the discipline in a new way: to offer a visual record of everyday "native" life. His *Nanook of the North* (1922) explored the rapidly vanishing "traditional" Inuit practices of walrus hunting, trading, shelter construction and family relations—even if Flaherty was later criticized for intervening and reconstructing in a myriad of unrevealed ways (Rothman 1997, ch. 1). In 1974, ethnographer and ethnographic filmmaker Margaret Mead published "Visual Anthropology in a Discipline of Words," militating for film, and categorically rejecting excuses for avoiding moving images. She lamented the long-held belief that anthropology should mainly rely on elders' words as they recalled the old ways. She dismissed the often-repeated claim that film takes so much skill that it could not be widely taught. And she refused to cede the field because of cost, saying astronomers and physicists hadn't abandoned their instruments because telescopes and accelerators did not come cheap (Mead 1974). Much of this part of Mead's argument resonates even more strongly today—who can argue that cost and work are insuperable obstacles in the era of miniature digital cameras and nonlinear editing?

Mead went on to argue that filmmaking promised a hugely more effective form of *documentation*, a dense record appropriate to a discipline of density: "for the illumination of future generations of human scientists" (Mead 1974, 4). Now this promise of an ideal archive no longer carries much weight in the interpretive social sciences—such a value-neutral observation protocol holds no more sway in anthropology than it does in history or sociology. And yet the idea of an *excess* in documentary film remains, a surplus captured that might be unintended, even unnoticed in the filming.

For all her radicality in lobbying for visual anthropology, Mead's razor-sharp separation of documentation from interpretation does not seize the ambition of late twentieth- and early twenty-first-century documentary and ethnographic filmmakers—and certainly did not rise to the ambition of the Geertzian ambition of narrating a culture through specificity. What survives from the stance of Mead or Gregory Bateson, and very vibrantly in the last decades of observational cinema, is a remarkable reengagement with some aspects of early cinema. Hollywood habituated audiences to privileged camera angles—impossible places for the camera, such as shooting from behind a fireplace, through the flames into the room; cameras shot as if through walls, mirrors or ceilings. By contrast, mass market films taught viewers to accept ever shorter shot lengths, along with a myriad of other conventions, such as match cuts (a visual similarity) that simulated continuity. As MacDougall (1998, 202) notes, "Implicit in a camera style is a theory of knowledge"— and the sudden changes of shot, match cuts and privileged angles vest knowledge in the filmmakers, who were after "essence."

MacDougall's own, more self-abnegating style of no-style ascribed a finite, observational knowledge to the maker, letting shots run long, filming from announced and physically possible positions. He began subtitling "native" speakers, instead of mute action interpreted by an all-knowing narrator; and his characters gained specificity and locality through real names and the registration of their own (subtitled, translated) words—for example, in MacDougall's *Wedding Camels*. At least some ethnographic filmmakers from the 1960s began putting the viewer in the position of a mortal filmmaker—not looking over the shoulder or listening at the feet of an omniscient observer (MacDougall 1998 204–205; Grimshaw and Ravetz 2009).

Documentary can put the observational process itself under pressure, as Errol Morris does in *The Thin Blue Line* (1988), where, through film-based recreations, he shows that the police account of a murder could not have taken place. In Morris's film, cinema *makes* the argument; the documentary becomes itself a generator of forensic evidence, and in fact reshaped a capital case. One can imagine film recreations serving an analogous role in STS. Suppose, for example, we wanted to know what impeded reproduction of a Transversely Excited Atmospheric (TEA) laser—was it the failure of print to capture what would be present in a moving, visual record? Was it something in the tone or gestures of the original designers that never found its way into documentation? Was it the ability of the replicators to ask follow-on questions? Was it some ineffable presence in the laboratory, a kind of unspoken confidence that the thing would work? With film, one could begin to explore—possibly eliminate, or in any case narrow—the spectrum of possibilities.

Other recent films open the observational eye to foreground the often-hidden interaction of scholar and subject. Anthropologist Stéphane Breton's *Them and Me* (2001) opens with our seeing Breton's camera-eye view as he walked down a rainforest path, following and filming a group of Guineans. Suddenly, one of the highlanders turns and says to the camera, "Do you like bananas?" From that moment on, we are sometimes painfully aware of the anthropologist—Breton offering work, paying the locals and endlessly negotiating, sometimes jokingly, sometimes tensely, with the highlanders for position and authority. Locality is here embraced by the investigator, not so much by reflexivity at the end (as in *Chronicle*) as by inclusion.

One is reminded of MacDougall's comment that film style brings a theory of knowledge—and each of these moves does just that: the film-maker-directed explorative recreations, the filmmaker's self-implication in the events observed, the self-conscious importation of stylized elements of transitions, the long shot rather than quick takes. Each choice surfaces the nature of knowledge production—in objects, among people and for us as we struggle to understand.

Film excels in depicting locality, materiality, scale, affect. Film plays with duration and simultaneity, setting people and work into immediate context, making landscape and built environment into characters, not described once but persistently present. Film reinserts a factory worker at the lathe among the machinery in each frame—not once, as in print, but over and again. Film can capture the simultaneity of reactions in a group shot, the scale and scope of a world as it acts on those in it. Film, as a time-based medium, conveys duration of work processes, permits simultaneity of appearance and action and mixes argument and affect. Returning to Goldmann's anti-*Chronicle* screed (claiming film fails as scholarship because it is too local and therefore insufficiently conceptual), one thinks of Kant, whose *Critique of Pure Reason* (A51/B75) was precisely about the fundamental relation between intuition (perception) and understanding: "Thoughts without content are empty, intuitions without concepts are blind. The understanding can intuit nothing, the senses can think nothing. Only through their union can knowledge arise."[8] When we pit conceptual abstraction *against* sensory particularity, we do so at our peril.

WHAT COULD VISUAL STS BE?

In a certain first-order sense, STS already knows the start of the argument I am making here: For the last thirty years, we have explored in a myriad of ways how concepts, algebra and text have never exhausted how science works. *Images*—diagrams, drawings, bubble chamber pictures, anatomical plates, astrophotographs, cartoons—and other elements of visual culture are inescapably part of science, irreducible to propositions alone. My argument is that at a second order, STS is also using and has the potential to use much more intensively images to establish its arguments—photographs of laboratory interiors, blueprints, patents and instruments. The interpretation of a notebook page, the gloss of a lithograph of a skull, the analysis of watercolor of a cloud study—these are using visual materials as constitutive parts of a scholarly argument, and not just as decorative illustrations (e.g., when writing about Joseph Fourier, inserting an otherwise uncommented-upon portrait).

Without doubt, VSTS can, like visual anthropology, capture process. For the visual anthropologist, building an igloo, preparing a dowry, herding sheep is important—for an STS attentive to science in the making as well as its finished products, this is crucial. We are, after all, very concerned with the creation and transfer of skills—skill in micropipetting, tracking animals, building particle detectors or analyzing data. Indeed, some of the most interesting claims of STS build on practices—the trained, skilled set of tools is one way to show the location of specific scientific work in a broader technical and even nontechnical world. Think of Joule's work on heat, which borrowed so powerfully from the stirring, insulating, temperature-measuring processes of his father's brewery. Otto Sibum has effectively used recreated instruments to get at materiality and work: Joule's, for example (Sibum 1995).

Let's dig down deeper into the *first order* (the study of scientific images that were mustered by the scientists to make their arguments) and second order (the use of a visually structured argument in the STS approach itself). Of course the two have no hard and fast boundary, but the difference in emphasis, at the limits, is clear.

First order: We could already see a role for a visual STS in the extraordinary outpouring of work *about* images and objects. These include an embrace of visual sources, not only diagrams but also photographs, cartoons, x-rays, drawings and simulated images. Out of this work, the STS constellation of fields bears directly on scientific epistemology. For example, we have a growing literature on diagrams—think of Bruno Latour's (1990) portrayal of diagrams as "immutable mobiles," sliding across time and space far more easily than words. Conversely, David Kaiser (2005) showed how *differently*, in fact, the "same" Feynman diagram worked in physics departments in Pasadena (Caltech), Berkeley (University of California) and Cambridge, Massachusetts (Harvard).[9] Such deepening inquiries into the practical, *differentiated* epistemology of diagrams are important: They show how central the visual can be to the shared toolkit of practitioners—to the doing of scientific work. Indeed, image making, image modification and image use constitute a form of epistemology—debates over the evidentiary status of statistical data (assembled by electronic counters) as against "golden events" (recorded by photography) cut through many branches of fundamental physics (Galison 1987, 1997).

We have begun to reckon with the variety of ways by which images enter scientific practice. Jennifer Tucker (2005), for example, showed how swiftly photography joined science in Victorian England—as a form of virtually witnessed evidence, from psychical research to astronomy, balloon expeditions and botany. How images can shape the very formation of a new scientific field—such as the functional nuclear magnetic resonance images that have made the field complex of "neuro-X" so immediately recognizable (Dumit 2003). Think of the myriad functions that the visual has played in Victorian geological field sketches, maps and even cartoons (Rudwick 1988); or, as Hanna Landecker (2007), has shown, in the study of microscope-based films of cellular biology; the ways aerial photography reshaped mid-twentieth century urbanism (Haffner, 2013); or in the exceptionally interesting contrast Michael Lynch and John Law (1988, 1999) make between the uses of "impoverished" (but more effective) drawn bird figures and the "realistic" (but overdetailed) photographs.[10]

Hanna Shell's (2012) examination of filmic treatments of camouflage shows the chameleon-like function of visual disguise both for the military (snipers) and biology (adaptive coloration); Marga Vicedo's (2009) deconstruction of Konrad Lorenz's "mother imprinting" films functioned on many levels—mother duck Lorenz led ducklings, with film both as his witness and evidence. Alongside both, film helped him reinscribe himself from an important, Nazi-sympathizing ethologist to a leading cultural icon of postwar Europe. The evidentiary, witnessing and popularizing functions of photography and film are crucial in helping us understand the nature of scientific practice from the Victorian era to the present. David Kirby (2011), meanwhile, tracked the shifting, sometimes unstable role scientists have held advising Hollywood productions—these advisors offer clues to creating a "reality effect" of laboratory or field—while working under the genre, plot arc and aesthetic choices that directors and audiences expect.

Nature and science films not only convey evidence and reposition the scientist, as Gregg Mitman (2009) has demonstrated, but also carry implicit scientific politics. He points (for example) to James Algar's *Nature's Half Acre* (1951), which was distributed through Disney. That Academy Award–winning film, with its anthropomorphic privileging of natural harmony (and bourgeois values) above evolution, celebrated and humanized nature, while leaving Darwin (random variation, selective retention) entirely on the cutting-room floor. If Disney emplaced its scientific-political ideology in the birds, scientific film could also be explicitly normative. With the stakes high, several social scientists of the 1930s wanted to use the medium to bolster democracy for a generation growing up in the shadow of fascism and Stalinism. As Javier Lezaun (2011) has shown, the child psychologist Kurt Lewin took this view, as he strove to establish a laboratory-instantiated, film-captured picture of democracy. "If Science," Lewin wrote, "is going to help to establish the reality of democracy for the young American it cannot be a science dealing with words. It will have to be a science dealing with facts; with facts of a very tangible nature; with facts close to the everyday of the individual person; with facts that matter."[11] Though he addresses far different topics (in Indian art, religion and politics), Christopher Pinney (2004) is after something similar when he suggests we need a "corpothetics" that directly implicates the viewer (as opposed to Enlightenment "aesthetics" that treats the viewer as distant, irrelevant).

Now turn to what might be labeled second-order use of image making—not images as the subject matter of inquiry, but instead images as a form, in its own right, of generating knowledge about the practice and place of science, technology and medicine. A few examples can illustrate ways that a VSTS filmmaking might complement print STS. To highlight the particularity of the visual, it is perhaps worth focusing on a few pairs of books and films—using documentaries not made to be part of a visual STS, yet highly suggestive of the kind of work such visual inquiries could do. Consider Frederick Wiseman's wry *Primate* (1974, fig. 9.1) alongside Donna Haraway's *Primate Visions* (1989). Haraway's study functions in several registers through several sites: It is a history of race, class and gender; it is a story of the Museum of Natural History in New York City—but also a narrative of colonial Africa, focused on Nairobi. It is about the ideology of big game hunting; about domestic interactions between the naturalist/hunter Carl Akeley and his two very different wives; and about "reading" the Museum of Natural History. Print easily flips between scenes and spaces, shooting elephants with bullets and film. *Primate Visions* can juxtapose the rhetoric of "manly triumph" vividly against "cultural decline."

Wiseman's film is confined to a single institution (Yerkes Primate Observatory in Atlanta), and it gathers its force by a fierce observational attentiveness to this one site—from technicians' artificial insemination of a primate through a young woman technician cradling and testing a diapered young animal, to the vicissitudes of the older apes, and eventually to experimentation on great apes, followed by their dissection and discard. The apes' vocalizations pierce the soundscape—so does the incessant clanging of steel-barred cages. Throughout, the camera holds the regard of the great apes, from newborns to 400-pound adults. Having witnessed these scenes in Wiseman's film, no viewer would see this labwork merely as a dismissible path to a crucial conclusion. Unnarrated, in black and white, the camera cuts back and forth between humans and apes, alongside sometimes disquieting, sometimes hilarious parallels, alongside measured, visceral violence. Trapped in the routine sounds and scenes of Yerkes, one learns

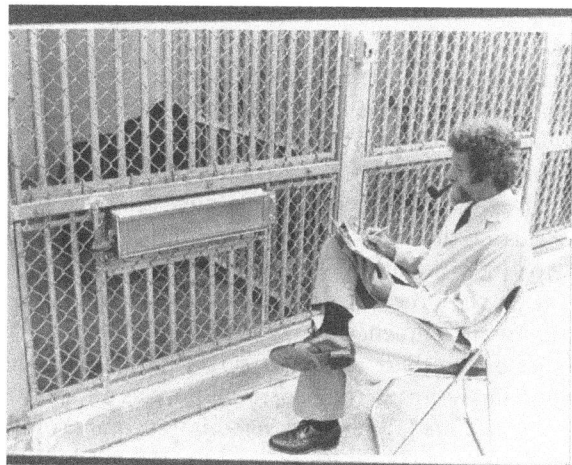

Figure 9.1   Still from *Primate*, directed by Frederick Wiseman (Zipporah Films, Inc., 1974) 16mm, 105 min., black & white. Photo provided courtesy of Zipporah Films, Inc. © 1974 Zipporah Films, Inc. All Rights Reserved.

about proximate not distal context; one sees the local production of science *in its place*—even if the scientific goals are never explained. Gender, power and interspecies relations in the laboratory are performed; they are thrown onto the screen—but *not* explained under an explicit conceptual frame grounded elsewhere. It is a very different, if complementary form of access to science-as-process than the equally rich, many-fold excursions of Haraway's reading of histories and symbols.

One could similarly contrast some of Wiseman's other, institution-based essay films: *Hospital* (1969), with its routine surgeries, waiting rooms and inner-city turbulence, might well be compared with Foucault's ([1963] 1973) *The Birth of the Clinic*. Both focus on the physician's encounter with the patient—the medicalization of interaction. But their forms of knowing differ; while Foucault attends to the radical structural shift in the "medical gaze" before and after the late eighteenth century, Wiseman's world is messier, more closely tied to circumstance: medicine in the midst of poverty, addiction and bureaucracy, and medicine on the phone, in the waiting room, and in cardiac surgery. Similarly, his *Missile* (1987) brings workspace interactions front and center: The film follows ICBM launch controllers at the 4315th Training Squadron of the Strategic Air Command at Vandenberg Air Force Base. Here are everyday rehearsals for entering codes, securing the control room and lofting the missile toward Moscow or Beijing. In one scene members of the 4315th celebrate receiving certificates for good performance (rapid launch); in another, yellow school buses ply their way around the base—in these small, extended moments a very different contextual awareness emerges than from analytic works such as MacKenzie's (1993) *Inventing Accuracy*, where the author shows how 1980s first-strike global strategic thinking shaped seemingly "pure technical" reasoning about accuracy. Wiseman and MacKenzie localize missiles differently. Wiseman shows us the ordinariness of what it looks and feels like to learn the movement, callouts, switch flips and key-turns that will end a million lives. MacKenzie shows us how the gyroscopes on top of those missiles carried their first-strike design in their very being. We might see these as two kinds of truth that complement each other: textural on one side, structural on the other. If we ask how the technology of intercontinental ballistic missile launches works, one would be hard-pressed to say (Goldmann) that one of these responses is truth, the other fiction.

Stephen Hawking writes in the register of physics and popular science—his *Brief History of Time* (1988), with ten million copies sold, stands as the world's best-selling popularization of science. Hélène Mialet (2003) uses STS tools to bring Hawking out of the mythological clouds, showing that the physicist exists not just as an isolated individual but also through assistants, technologies, students and nurses. By contrast, Errol Morris in *his Brief History of Time* (1991) covered Hawking through a highly produced character sketch that moves among stage sets, theory, images and biography. Taken together, the Hawking and Mialet texts and the Morris film show *different* contexts—Hawking presents a popularized, synthetic version of results in quantum gravity; Morris follows a filmic logic of association and sharp-focused, psychologically dense materiality; and Mialet peels back the layers of a decentered figure.

Morris enthusiastically uses high production values—drawing on a long career in the production of vivid, precise television advertisements—to depart from Wiseman's rough-hewn verité style.[12] Michel Negroponte chooses a third camera style, neither verité nor staged. In his expressive, impressionistic film (*I'm Dangerous with Love*, 2009, fig. 9.2) Negroponte follows Dimitri Mugianis, former musician, now dispensing an illegal hallucinogen, ibogaine, through the underground world of heroin addicts trying to shake their habit. Negroponte's is a much wilder, more expressionistic film than Wiseman's or Morris's. *Dangerous* certainly is *not* a white paper report about scientific institutions or clinical trials—instead, it is a highly subjective excursion into the improvised administration of a dangerous Schedule 1 drug, recently turned to a dramatically new use. Hope and panic, even near-death experiences of desperate people populate the screen, and the camerawork is correspondingly often handheld, not to show (verité) authenticity of grainy black-and-white footage but to capture the feeling of situations edging out of control. Along the way, we follow Mugianis's therapeutic trips in North America, his Bwiti shamanistic initiation in Gabon and his awkward import of that experience back into the margins of America. How different this is from Jeremy Greene's (2007) analytic work showing how a commercial blood-pressure drug, Diuril, came into existence accidentally—designed as a diuretic by Merck, Sharpe & Dohme. Once the company understood the antihypertensive properties of their product, its marketing teamed created an unprecedented drive on doctors using journal advertisements, visits from markers, orchestrated journal articles and the soon famous model, the "Diuril Man" that stood on physicians' desks from coast to coast. Here the complementary nature of film and print are laid out even more strikingly—through different kinds of locality: documentary analysis of advertising firms, their market plans and Big Pharma, but also images of a shamanistic initiation and a drug regimen administered on the fly, about to spin out of control. Drugs permeate our biological condition today, licit and illicit, high-tech/high-capitalization and street corner hustle, observational-expressionist cinema and analytic writing. STS needs both kinds of understanding; print and camera position each offer their own range of epistemologies.

In the postverité world, the long or immersed or reflexive camera positions each bring their own form of knowledge. But the digital age has begun to expand that repertoire even farther as camera size and location are no longer limited by proximity to the eye of the shooter. Nowhere is this illustrated more strikingly than in Lucien Castaing-Taylor and Véréna Paravel's *Leviathan* (2012, fig. 9.3). It is, I suppose, about commercial fishing off New England, but the point of view has shifted as the camera moves from the shoulder to a very small waterproof cam at the end of a pole. Sometimes the viewer is sloshed over by dead fish as they slide back and forth with the

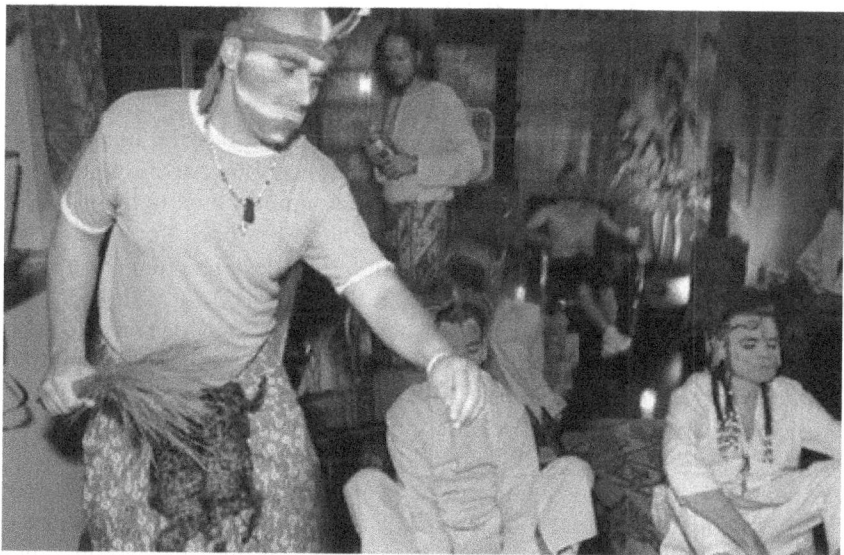

Figure 9.2 Dimitri Mugianis, photograph by Ashley Valmere, still from *I'm Dangerous with Love*, directed by Michel Negroponte (Blackbridge Productions, 2010), DVD, 82 min.

Figure 9.3   Still from *Leviathan*, directed by Véréna Paravel and Lucien Castaing-Taylor (Arrête ton Cinéma, 2012), DVD, 87 min.

pitch of the boat. Sometimes we move along the surface of the water by a run-off spout spewing a mix of seawater and fish blood. Sometimes we are below water altogether as a mix of fish parts, netting and starfish flicker by. Where are we epistemologically? Not so much in the Latour-Callon world of actants—scallops here, like people, nets and birds, are washing by, more in flux than networked nodes. This cam-eye perspective is a point of view that was just a few years ago technically impossible; but it is now more than an electronic alteration. Here in the visceral, off-human perspective is a shift away from character and broad explanatory context—and toward a carefully constructed, sensual, horrifying, dizzying moving image. Is it knowledge? Certainly. Is it subordinated to a conceptual partition, an economic or actor-network explanation of the economics, biography or regulatory structure of fishing in the early twenty-first century? Not at all.

Film and texts here each facilitate and impede particular forms of understanding.[13] Both recreate worlds, but they are very different. Greene, MacKenzie, Mialet and Haraway cut across spaces and registers to provide conceptual-explanatory schemes. Wiseman, Morris, Negroponte, and Castaing-Taylor and Paravel, by contrast, use different means (observational juxtapositions, recreations, immersive, off-human filming), and together offer a probing phenomenological vision of the sci-tech world. Is text contextual and film isolated? Not at all: Text and film each pick out *different* contexts, textual and textural. It seems to me a bit like imaging a distant star: Look with a radio telescope, an optical telescope and a gamma ray telescope, and each could be said to simultaneously obscure and reveal. Together, print and visual STS could offer a more dimensional, denser understanding of the world of science, technology and medicine.

FILMING AND WRITING SCIENCE

Now a hard cut to a more personal point of view, to the back and forth trajectory I have taken, between filming and writing. When I was writing *How Experiments End* in the early 1980s, images entered in three ways. First, as a matter of content, I was struck that some of the experimenters I was studying relied so heavily on visual forms of evidence—I was immensely struck by the way a cloud chamber image or two (like the one that persuaded Carl Anderson of the reality of the positron) could have such persuasive force. By the time I finished the book, it had become clear to me that experimenters using cloud chambers or bubble chambers, for example, had a very different working epistemology than those immersed in counters and their associated statistics. This evidentiary contrast later became a guiding theme of *Image and Logic*. Second, I had in mind a role for images beyond the mounting of an argument. As an exploratory move, I wanted the images as a whole to form a counterpoint argument of *How Experiments End*. That is, I wanted the reader to be able to read the book two ways: with text and image—or on a different plane through images and captions alone, a kind of parallel, flip-book story among machines, evidence, diagrams and architecture. The idea was to make a kind of graphic narrative, one that would use photographs, diagrams and other imagery to tell the book's story, but in a way that would complement the main text. My hope was that as the reader went from devices the size of a table to that of a factory, I wanted the scale shift to register in more than words.[14]

Around the time I was writing *How Experiments End* (in the early 1980s), I was also testing to see if I could push a level down, to the development of instruments—hydrogen bubble chambers, for example. One day, I was wandering around the remains of an old hydrogen liquefaction plant in Denver with one of its creators and he remarked that the bubble chamber hydrogen supply had come ready to go from work on the . . . [mumble]. I asked again. Mumble. Finally, more clearly: "It came from the nuclear device on Eniwetok," the site of the first (liquid) hydrogen bomb. I went back to the archives, where, buried in a box of purchase orders, I saw the packing forms that had sent the liquefier from the Pacific to Berkeley, where it became a workhorse of one of the most successful campaigns of particle physics ever conducted. Working backwards, I began to see how so much of the postwar physics boom began in the surplus equipment, novel forms of organization and technical skills developed in the World and Cold Wars.[15]

I wanted to know too how the physicists understood themselves—two questions fundamentally interleaved: What did it mean to be a physicist, what counted as physics in 1920, 1950 or 1980? To know one, one had to know the other. I began talking quite a lot with Luis Alvarez at Berkeley, who had played such a large role in World War II radar and nuclear projects—and went on to be a key figure in the Cold War. I spoke with Hans Bethe, Edward Teller, too, and read extensively in their and others' archival papers. What struck me about the early Cold War—from 1947 to 1952—was how volatile positions were toward the nuclear ramp-up that followed Hiroshima and Nagasaki. It first seemed hopeless to find an order; toward the hydrogen bomb one saw with so many of the figures what seemed a random flipping: for it, against, it, for it, against it, resigned to it. With Barton Bernstein, we began trying to sort this morass of moral-political positions—and eventually I came to see that it was far from an uninterpretable mess. Instead, it was rather sharply periodized by historical inflection points (Galison and Bernstein 1989). A general acceptance during World War II; a turn against new nuclear weapons after the human effects of Japanese bombings; an acceptance that H-bombs should be a part of the package of nuclear arsenal in the period after the start of the Cold War, and a profound polarization following the Russian detonation of their first bomb, "Joe 1," in the summer of 1949.

It was then in the early 1980s that I saw John Else's *Day after Trinity* (1981), which I found electrifying. Here was a film about science and technology that was not a glorification of the atomic bomb, a diatribe against armaments or a ginned-up story of the race to build it, but instead an exploration of J. Robert Oppenheimer—how in the context of the war, he went from being a somewhat shy, philosophically inclined physicist to the leader of a massive effort to build the atomic bomb. That film opened a world to me—for the first time, I began to think about how I could merge my interest in filmmaking with the kind of science studies that intrigued me. I began to imagine filming people I knew—Hans Bethe, for example—alongside archival and other visual footage to try to get, while it was still possible, a film account of the *moral-political* history of the hydrogen bomb.

Making *Ultimate Weapon* took my collaborator, Pamela Hogan, and me a good long time—more than a decade. We first worked on video through the dreadful, "portable" system of analogue videotape—a camera (a Sony, I believe) linked by cable to a half-inch videotape recorder. Though at the time, the fairly heavy cameras could be carried (not just put on a tripod), editing—*linear* editing—meant that in order to insert a section, call it C, on one tape between shots A and B (located on another tape in the order AB), you would need to first put the tape with shots AB on it into a video player, call it the source player, and copy just A to a fresh tape on the target recorder; second, put the tape with shot C on the source player and copy that after the copy of shot A on the target tape (now the target tape has shot sequence AC). Third, put your first tape with AB back on the source player and copy shot B onto the target tape after the portion of the tape with AC. Now you have on the target tape ACB, as desired, but this target tape is a generation down from your original A, B and C—more snow, more flickering. By the time you had done some serious editing, you had to squint, pray and imagine your way to decoding what was on your working tape. The most often copied pieces would be so many generations from the original that it looked like it was just in from Alpha Centauri.

Beyond technical horribleness, I had no real sense in the late 1970s and early 1980s what I could do with video filming. Science documentaries tended, on the whole, to be pedagogical or visual white papers. The topics (the story of DDT, for example) were important but not very cinematic: the race to X, the discovery of Y, the wonder of Z. It was a time when some of us—from history, sociology and philosophy—were beginning to think about science as process, not results; it was a time when the goal was not to celebrate or popularize discoveries but to explore science as a process.

Some of my Zeno-like progress was of practical origin (the nightmare of analogue video editing), but it was more than that. I had to *un*learn much of what I took for granted in writing. Take interviews: I had done hundreds of oral history interviews—they were crucial to so much of my work as I tried to bring history into the present. So I thought I was pretty good at it, and would just move the interview onto film. Wrong. In an interview for print, ellipses and fragmentary excerpts, contextualized in the surrounding sentence, are your friends. Ellipses (cutting away bits) are not such good pals on film. You can also conduct a print interview most anywhere: A fly lands on the forehead of your interview subject in a print interview and she swats it away—you don't even notice. Film it—and no one will see anything but the fly. I once hiked out to the middle of nowhere to film the pilot of the fallout-filter plane who, in 1949, had discovered that the Russians had detonated a bomb. I drove miles and miles in the Western desert sun, clambered up to his apartment over his grown children's garage, set up in the killing heat . . . and lost the interview because I forgot to turn off the refrigerator. It was humming, loudly and intermittently.

Worse still, I'd get affectively flat responses to questions, which didn't matter much for print. My question to a hugely creative figure at the intersection of computer design, math, physics and bomb design: "Was it surprising to have use of one of the first computers to calculate the force of the H-bomb?" Interviewee: "No, I wouldn't say exciting. But it gave good results." Pamela Hogan turned to me and said, "We came to Los Alamos to get this?" So I tried again. Me (approximately): "So, people think that putting a simulated weapon on the computer is pretty standard—is that right?" Interviewee: "Standard? That was the first time anyone had run that computer at all—and the first thing we ever put on it was the hydrogen bomb. It was anything but usual."

But worst of all—worse than learning to see, worse than juggling affect, content and scene simultaneously—were more structural problems. My first idea was to take the article Bernstein and I had written and put it into a script. There would be six stages:

1) World War II through Hiroshima.
2) End of World War II through the failure of the Baruch Plan in 1947.
3) Failure of Baruch Plan to the Russian bomb, "Joe 1," in 1949.

4) Joe 1 to Teller and Ulam's January 1950 sketch of a plausible H-bomb design.
5) Teller-Ulam idea through the first H-bomb test in late 1952.
6) Early Cold War, after Joe 1.

In print, this partition had put order into the apparent flip-flops—for example, once there was a plausible design that the physicists assumed the Soviets would soon find, many dissenters gave up.

Why *not* build a film that way? Because an illustrated lecture is not a film. This blatantly obvious fact was the hardest lesson of all. I'd worked so hard for so long to put order onto material, to make sense of a vast morass of archival and oral material, it was hard to give up. And my periodization chart that made sense of why Fermi (for example) changed his mind after Joe 1 . . . or why Bethe changed his after seeing the Teller-Ulam solution in January 1951. True, one can force film into written structures, use charts, divide into chapters with intertitles . . . but then the specificity of film is lost. Yes, you can rig a schooner with immense outboard engines to move faster in the water. But if speed is your goal, it is the wrong tool for the job. If you want to explain what a second cousin once removed is, use a kinship chart. If you want to make a point-by-point comparison of postmodern architecture of the 1980s with modernist architecture of the 1960s, print serves well. A very good book in STS, Steven Shapin and Simon Schaffer's *Leviathan and the Air-Pump*—one of my favorite works in the field—starts as follows: "Our subject is experiment. We want to understand the nature and status of experimental practices and their intellectual products."[16] That launched the book well—but it is inconceivable as a way to begin a film. Film needs to hit the ground in the material and particular.

What I slowly learned was that film did other things. In one scene from the H-bomb film, Theodore (Ted) Taylor, one of the hotshot young thermonuclear weapons designers of the 1960s and his wife, Cara, were talking to each other about explosions. Ted: "Big explosions, quite aside from making weapons, people being killed, big bangs, I've always loved them." He then turns to Cara, who is looking at him with a mixture of affection and utter incredulity. "But," she says softly, "you don't go to the fireworks to hear the bang." Ted: "Well, I did." That exchange of looks, words and body position, the density of the interaction—it was a quip in a lifelong conversation.

Film could also contain images, both moving and still, that were themselves part of the contested nuclear era. Oppenheimer, General Groves and their colleagues actually starred in a recreation of the race to the bomb—we could use those performances. The mission to detonate the first hydrogen bomb was filmed and projected by the Atomic Energy Commission (AEC) for the relevant congressional committees. Visual material that might seem to be purely a record was more than that; it was an integral part of the historical production of the thermonuclear weapon. We could and did use some of those visual fragments too, repurposing them. Indeed, this reuse of material to different ends is fast becoming an important feature of eth-nographic and documentary filmmaking today. Gregg Mitman and Sarita Siegel, for example, have found and digitized anthropological footage shot in Liberia in the 1920s, in the boom years of Firestone's rubber exploi-tation. They are, at the time of this writing, in the process of retracing that expedition, showing the old clips at villages that were stopping points on the trajectory taken by the Western anthropologists ninety years ago (and filming the encounters). Mitman and Siegel have found that people take these "archival" film segments up in ways far different from anything expected by the original makers. The clips become part of a reconstruction of history, much of which was shattered in the long and brutal Liberian civil war. At the same time for individuals, some of whom are seeing their direct relatives, the film and film viewing become part of a more personal, familial reassessment. Out of this back and forth between history and con-temporary, reciprocal anthropology, the director/authors are making a film (*Where the Cotton Tree Grows*) that is, all at once, about the circulation of goods, like rubber, the migration of diseases and the fragmentary rebuild-ing of familial and national history. Could this be done with print? Hardly. *Reading* a 1926 journal to a town meeting in Liberia simply would not have the density, affective, personal or historical, of this rediscovered and repurposed ethnographic film.

Getting back to H-bombs—my film with Pamela Hogan, *Ultimate Weapon: The H-bomb Dilemma*, came out in 2000. It was then that I began collaborating with Robb Moss, whose films (e.g., *The Same River Twice* [2003]) use observation to record the personal, in very innovative ways. While working on the intersection of the history of physics and the development of nuclear weapons, I had begun thinking about the dynamics of classification. On a purely analytic level, secrecy was important because it seemed to me a kind of "anti-epistemology." To come up with a process that would block transmission, was, in effect, to show what your theory of knowledge was. I began writing about this negative account of knowledge (Galison 2004). But secrecy is more than epistemology—in our conflict-riven world, too much secrecy about political and military matters is a threat to the very possibility of deliberative democracy. At the same time, secrecy is never *just* a matter of procedure—and I was intrigued by our society-wide psychological fascination with forbidden knowledge. Think of the American congressmen who prized secret (but false) reports of Iraqi nuclear weapons above open (as it turned out, verifiable) reports that no such rebuilt weapons program existed.

Against reason, Robb Moss and I began thinking about how we could make a film about national security secrecy (fig. 9.4)—against reason because if one listed filmable topics from the most obviously visual to the least, I'm not sure what would be farther *down* the list than secrets: the things that could be neither spoken nor shown. Yet even this idea—filming the hidden—intrigued us. But the style would have to break with that of

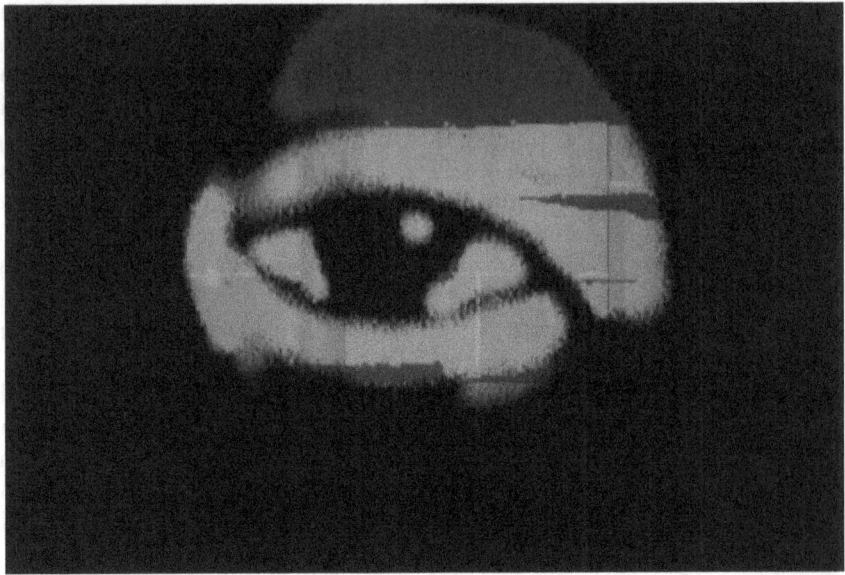

Figure 9.4   Still from *Secrecy*, directed by Peter Galison and Robb Moss (Redacted Pictures, 2008), DVD, 85 min.

Ultimate Weapon in a variety of ways. We set ourselves some rules from the beginning: no narrator, no pundits, but instead people—analysts, interrogators, journalists, citizens—all caught up in the system of secret making and keeping. Above all, we had to figure out a way to convey the part of secrecy that went beyond the rules of classification. We wanted also to convey the affect that surrounded secrecy. What did people want from secrecy, what effects did it have on those inside—and outside?

Here is an instance that did not work. Early in the project, Robb Moss and I went down to Maryland to film a Department of Energy official, who had played a central role in protecting secrecy around the nuclear arsenal. He had many important things to say, including his and many of his colleagues' view that *over*classification, rather than putting a cordon sanitaire around the deepest secrets, actually degraded secrecy by abusing it. So there we were, shooting as this fellow moved around his Chesapeake Bay waterfront, wavelets lapping on sand, wind gently swaying the leaves in the trees. *Nothing* about this scene had anything to do with secrecy; indeed, the whole situation radiated the *opposite* of secrecy.

This called for a major shift; it was then that we began sketching out the idea of a carefully controlled sound stage of light and dark and an aesthetic that would make the theme of visibility and obscurity, and the shifting flows of information. In this period (2003–2007), digital tools advanced far beyond those of the all-too analogue editing of *Ultimate Weapon*—both storage and processing speeds made possible immediate access to a great deal of footage, cut, fade, slow down, in quick succession. Take slowing-down: In analogue, either one films a scene with specialized high-speed equipment and then projects it at normal speed—or one laboriously inserts copies after each frame (frames ABC become AAA, BBB, CCC, etc.) Such laborious, expensive work is now an easy and routine component of any digital editing suite. A more elaborate and fluid effect interpolates between the images on one frame and its successor, a feat simply impossible before digital. We used this effect (combined with increasing the contrast of the original film clips from World War II) in slowing down a short segment of Japanese planes diving toward Pearl Harbor—we wanted our black and white, high-contrast xeroxes of documents to be visually echoed in some of the moving sequences.[17]

Documents were central to secrecy and to our story of its dynamics. Our first secrecy episode (of three) was about the legal case that established the State Secrets Privilege, *Reynolds v. the United States*. Back in the early 1950s, the family of a civilian engineer killed in an Air Force crash sued the government to get the accident report. The government refused, and in 1953 the Supreme Court backed the withholding of evidence. The justices themselves would not read the contested document. When the family discovered the crash report years later, they found that, in fact, no secrets had been in the report. By then it was too late—the State Secrets Privilege had become a pillar of national security secrecy, a kind of superprecedent for a myriad of other cases. The fruit of that poisoned tree had grown into a full-blown toxic orchard.

Film could convey, more than print, the life-altering effects this blacked-out information had had on the wives and children of the men killed. In the spaces of censorship, we spin theories. To my surprise, secrets were never just routine, bureaucratic business, even to those who work with them every day. When a senior person at the National Security Agency said, "Secrecy is like forbidden fruit, you can't have it . . . makes you want it more," he was not alone in invoking biblical-sexual knowledge. Almost every person in the film (those for more as well as those for less secrecy) moved fluidly between matters of state and personal secrets. To capture this overflow of associations we enlisted an animator, whose German-expressionist-like animations, white lines on black background, formed a kind of unconscious of the film, an extension of arguments when the words no longer sufficed. In filmed footage, American soldiers break through to an underground bunker in Iraq, and begin rooting through rooms and crates—the play of a flashlight dissolving into a digitally manipulated, animated sequence of what they hoped to find. Later in the film, a former CIA interrogator, Melissa Mahle, described on camera the impact of her years of deception on her relation with her mother; she spoke of the effect of having her father pay for a sham wedding, when the real one had to be held in secret. Images in the sequence shift from wedding photos to a gestalt-shifting animation of two figures, faces to a silencing finger held vertically. Such imagery can carry implicitly a density of associations.

Robb Moss and I are now in the midst of another project: *Containment*, about nuclear waste. In several hundred tanks, each the size of the Capitol dome, sit staggering amounts of radioactive waste, the detritus of making or refashioning seventy thousand nuclear warheads. This high-level waste has the consistency of peanut butter; some tanks are still boiling, while others are in danger of exploding from the hydrogen gas "burps" emerging from their depths. It would be fatal to approach the outside of a tank if it were above ground. Yet not a single country has a fully worked-out plan of how to handle this reprocessed material. High-level waste is leaking into the groundwater and crawling into the Columbia River. Film—from digital cameras inside these million-gallon tanks—make visible these extraordinarily toxic sites inside inaccessible weapons factories.

Nuclear waste inescapably extends far over space and time (Macfarlane and Ewing 2006; Masco 2006). In terms of space, lands of the nuclear weapons and power establishment are bigger than some American states. The health problem of waste is one of migration through water, air and soil, and into living bodies. One side of the moral-political conundrum hinges on who lives near these sites—in many cases, Native Americans and African-Americans, or in other countries rural Pacific Islanders, Tibetans and Algerians.[18] In terms of time, because many fission and transuranic elements have long half-lives (plutonium's is 24,100 years), there is a second moral-political issue—nuclear waste will affect people for ten half-lives (a quarter of a million years), raising the question of intergenerational equity.

This twinned set of problems (material-present and imaginative-futurological) set the two broad parameters of *Containment*. We are using observational filmmaking to capture the texture of adjacent communities and the industrial plants of weapons, power and burial sites in the nuclear waste complex. But nuclear waste is with us for the future *longue dureé*. So the film must also follow the people (astronomers, science fiction writers, semioticians, anthropologists) whom Congress and the Department of Energy asked to imagine scenarios of inadvertent intrusion into the waste sites, ten thousand years from 1989. Another team would then use these envisioned blunders to design monuments that would deter our descendants from killing themselves by exposure to the buried isotopes. This warning structure prompted us to join interviews and observational film with excursions into animation and simulation. It is a way of capturing the dual nature of radiological experience—*picturing* "scenarios" alongside the quotidian efforts of mining and political wrangling is a way to inhabit the ordinary physical and the range of fear and forecast that always accompanies the big atomic projects. Nuclear waste, like national security secrecy, is precisely the kind of topic that VSTS can effectively address. For here lie arenas where understanding can be advanced by visualizing otherwise hidden aspects of science, the state and contemporary life—with allusions and materialities riding side by side.

Just a few years after *Secrecy*, there are more useable digital tools—for depicting a consequential but imagined far future. We can use 3-D-animation to bring planned monuments into fully visual forms. We can take a graphic narrative of a scenario and insert limited animation into one or two elements, as a way of registering the incompleteness of these schematics of future societies. These elements fill out the world of nuclear waste in a way that print cannot—because the world of radioactivity, risk and disposal is never just about statistics. For better or worse, the fate of nuclear power, weapons and waste is bound up with our shifting collective understanding of our obligations to the far future. Knowing, *picturing* what lies behind the fence, in the tank, under the ground, matters.

CONCLUSION: TRUTH IN THE DISTAL, TRUTH IN THE IMMEDIATE

Some of the most interesting work in STS these days cuts across the pure disciplinary categories of history, philosophy, sociology and anthropology. No doubt that is and will be ever more true in VSTS. Borrowing from the sophisticated and interesting tradition of visual anthropology (and the more recent field of visual studies) would be a good idea—I have tried to point to some of the remarkable work that has marked that line of visual achievements. I hope at the same time that three features of STS itself can continue into the heartland of what is a burgeoning field.

STS has in some of its strongest work taken on the knowledge constituting standards that form and define fields of scientific and social scientific work. What constitutes a proof, an experiment, an observation? What counts as a proper form of knowledge circulation or certification? How does one become a qualified scientist—how are clashes resolved or demonstrations brought to a close? STS has already made a powerful mark by exploring the mix of cognitive, institutional/political and ethical considerations, asking about the history and trajectory of experiment, probability, curiosity, objectivity and quantification.

When Wiseman entered the Yerkes lab, he was not asking about the scientists' own questions—he very deliberately wanted to stand outside of their concerns, their arguments or their own account of why they were there. That distanced stance led to enormous insight, alongside extremely funny visual and comportment parallels between apes and humans. In transforming Hawking's own *Brief History of Time* into film, Morris essentially set aside Hawking's own constant return to the physics, and in its place set front and center an imaginative, visual biography of Hawking, from childhood (or its memory) into a kind of thought-governed future.

In my view, STS is at its best when it refuses any bright-line division between content and context—it shines precisely by asking about sociality and the science, about ethics alongside epistemology. I hope that we can use VSTS to push on these sorts of problems, to drive them further, and to forge new alliances with digital innovation, visual anthropology and visual studies. Some of the deep political problems of our time would benefit hugely if we could find ways to visualize them into the fabric of local concerns. Drones and data mining, bio-privacy and emerging diseases come to mind—to name just a few sites of concern, ones where drawing the hidden into the visible could make both an analytic and a societal intervention.

But in the end, reporting on a nascent subject matter, the hope is that these reflections will be seen as an invitation, not a defended territory with a new set of walls. The best moments of science and technology studies have come from moments of openness—new means to deepen our understanding of scientific practice in the world—from history, philosophy and sociology of science to gender, media and colonial inquiries. Visual STS should build on work already accomplished through STS studies of still and moving images as evidence. It should attend to what can be learned both from the history of visual anthropology and from that of documentary film, as well as from the steps now being taken in nonlinear presentations, and see how they might touch on the fields of science, technology and medicine. Still images and sound recordings, film and interactive web design, interdisciplinary locative digital projects, experimental and observational filmmaking—the possibilities are many. By making as well as writing, we can get at the embedded, material particularity of science in ways that will sharpen our understanding of science, technology and medicine as they change.

Sometimes, contra Goldmann, there is truth not just in abstraction but through local, human and material concepts concretized precisely through the sense-density of film and other visual media. We have the possibility of exploring new ground not by opposing STS but by developing an enriched VSTS—to convey with greater immediacy the scale, scope, duration and materiality of our scientific-technical world.

NOTES

1. On locality in Actor Network Theory (ANT), see, for example, Latour (1987); on scientific authorship's site specificity, see Galison and Biagioli (2003); on book history in the history of science and STS and its dependence on historically specific context, see Johns (1998); on the means of transmission itself as a form of local, material analysis, inflected by history, see Kittler (1999)'s analysis of everyday communicative devices like the typewriter or phonograph.

2. On locality in laboratory studies, see Galison (1987, 1997) and Shapin and Schaffer (1985); on exemplifying local studies of procedure with larger claims of historical epistemology, see, for example, Davidson (2001) and Daston and Galison (2007); on the local sociology of confirmation and repetition, part of the empirical program of relativism, see, for example, Collins (1981); on ethnomethodography and discourse analysis, see Garfinkel (1967) and Lynch (1993).

3. Lucien Goldmann, "Cinema and Sociology: Thoughts on *Chronique d'un été*," in *Anthropology, Reality, Cinema: The Films of Jean Rouch*, ed. Nick Eaton, trans. John Higgins, 64–66 (London: British Film Institute, 1979), 64.
4. Ibid., 65.
5. Ibid.
6. Thomas S. Kuhn, *The Structure of Scientific Revolutions* (Chicago: University of Chicago Press, 1962).
7. On Newman's work: http://www.indiana.edu/~rcapub/v29n1/alchemy. shtml. For a time lapse image of the Tree of Diana as it grows, see Newman: http://webapp1.dlib.indiana.edu/collections/newton/chemlab/webvids/ silvertree/silvertree_Large.mov. Last accessed 4 November 2013.
8. Immanuel Kant, *Critique of Pure Reason,* trans. and ed. by Marcus Weigelt (London and NY: Penguin, 2007), 86.
9. For good starting points on visualization in science, see, for example, the essays in Lynch and Woolgar (1990) and Biagioli (1999).
10. Michael Lynch and Steve Woolgar celebrated the turn from purely propositional knowledge to the visual, all while rightly cautioning back in 1988 against too easily supposing that images could travel independently of context. See their edited volume (1990).
11. Javier Lezaun, "Demo for Democracy," *Limn* 001, accessed September 9, 2012, http://limn.it/demo-for-democracy/.
12. Morris (2011) reflects on the complexity of photographic context, the meaning of "manipulation" and the political valence of both.
13. A particularly good statement of "what images want" (as compared with print) can be found in an interview with W. J. T. Mitchell from January 2001, in Dikovitskaya (2005, 238–257); there may be no Chomsky of images, but there is no surplus in pure text to compare with that of an image.
14. In 1988, the year after *How Experiments End*, I read with fascination the just then translated work by Klaus Theweleit: <u>*Männerphantasie*</u>, a study of the right-wing German *Freikorps* between the wars. Theweleit used images from Japanese anime to home photos to string an argument complementary to that of the text, about the relevance of the fearful body-armored militaristic culture of the 1920s to the much later present of his own relation to his own fascist father, and the broader culture of the mid-1970s.
15. See, for example, Galison (1988). The physical infrastructure of physics once it was inflected by war went from building equipment in the 1930s for $1,000 to imagining the construction of bubble chambers for $1 million just a few years later. This change in scope altered the very self-conception of physicists—how they fit into the world, what spaces the laboratories resembled, how they related to industry, to the military, to society in general. In my work, confrontation with all this came to a head in the 1980s—from the inside out, as I began to see the embedding of instrumentation into the wider world of technical, military and industrial apparatus. And outside in: The huge expansion of nuclear weapons at that time dominated discussion not only inside physics but also throughout international politics.
16. Steven Shapin and Simon Schaffer, *Leviathan and the Air-Pump: Hobbes, Boyle, and the Experimental Life* (Princeton: Princeton University Press, 1985), 3.
17. Aiming for that same "xerographic" visual quality, we used an effect that has its origin in a very old technique known in animation circles as "rotoscoping." Back in the 1910s, Dave and Max Fleischer built an apparatus to project individual film frames on a frosted glass where they could draw the image, quite famously in "Koko the Clown." This let them create a one-to-one illustrated version of a film sequence made with a real actor, which

formed the basis of a famous series begun in 1915, about situations conjured through drawing, *Out of the Inkwell*. Our interest in rotoscoping was not so much a fairy princess so much as an immense landscape of offices and laboratories. So we strapped a video camera on a library pushcart, and rolled it down MIT's iconic "infinite corridor." We then increased the contrast (digitally), printed out the individual frames into a stack of images and xeroxed that pile over and over until we had the stark look we wanted. We then rescanned this digital sequence of frames into the film, and looped it. The final product reads as if it were a kind of precomputer document in motion.

18. On nuclear environmental justice, see Schlosberg (2007), Shrader-Frechette (2002) and Stoffle and Evans (1992). STS-derived concerns enter in another way too: Addressing nuclear waste inevitably involves the establishment of *trading zones* (e.g., Galison 1997; Gorman 2010)—regions of limited coordination in the midst of global clashes. With nuclear waste, we, as a society, must grapple with fundamental conflicts of the myriad stakeholders, driven by conflicting goals of profit, military power, climate change and environmental justice—alongside more recent ethical concerns about intergenerational equity. By following the president's Blue Ribbon Commission as it tries to navigate its way among warring stakeholders, we have, in the making of *Containment*, a field laboratory for studying an STS trading zone in action.

REFERENCES

Banks, Marcus, and Howard Morphy. 1997. *Rethinking Visual Anthropology.* New Haven: Yale University Press.

Barbash, Isisa, and Lucien Taylor. 1997. *Cross-Cultural Filmmaking: A Handbook for Making Documentary and Ethnographic Films and Videos.* Berkeley: University of California Press.

Biagioli, Mario. 1999. *The Science Studies Reader.* New York: Taylor Francis.

Burdick, Anne, Johanna Drucker, Peter Lunenfeld, Todd Presner and Jeffrey Schnapp. 2012. *Digital Humanities.* Cambridge, MA: MIT Press.

Collins, Harry M. 1981. "'Son of Seven Sexes': The Social Destruction of a Physical Phenomenon." *Social Studies of Science* 11 (1): 33–62.

Davidson, Arnold. 2001. *The Emergence of Sexuality: Historical Epistemology and the Formation of Concepts.* Cambridge, MA: Harvard University Press.

Dikovitskaya, Margaret. 2005. *Visual Culture: A Study of the Visual after the Cultural Turn.* Cambridge, MA: MIT Press.

Dumit, Joseph. 2003. *Picturing Personhood: Brain Scans and Biomedical Identity.* Princeton, NJ: Princeton University Press.

Feyerabend, Paul. 1975. *Against Method.* London: New Left Books; Atlantic Highlands: Humanities Press.

Foucault, Michel. (1963) 1973. *The Birth of the Clinic: An Archaeology of Medical Perception.* Translated by Sheridan Smith. London: Tavistock.

Galison, Peter. 1987. *How Experiments End.* Chicago: University of Chicago Press.

Galison, Peter. 1988. "Physics between War and Peace." In *Science, Technology, and the Military*, vol. 1, edited by Everett Mendelsohn, M. Roe Smith and Peter Weingart, 47–86. Boston: Kluwer.

Galison, Peter. 1997. *Image and Logic: A Material Culture of Microphysics.* Chicago: University of Chicago Press.

Galison, Peter. 2004. "Removing Knowledge." *Critical Inquiry* 32: 229–243.

Galison, Peter. 2010. "Secrecy in Three Acts." *Social Research: An International Quarterly* 77 (3): 941–974.

Galison, Peter. Forthcoming. "Blacked-Out Spaces: Freud, Censorship and the Topography of Mind." *British Journal in the History of Science.*

Galison, Peter, and Barton Bernstein. 1989. "'In Any Light': Scientists and the Decision to Build the Hydrogen Bomb." *Historical Studies in the Physical and Biological Sciences* 19: 267–347.

Galison, Peter, and Mario Biagioli, eds. 2003. *Scientific Authorship: Credit and Intellectual Property in Science.* New York: Routledge.

Garfinkel, Harold. 1967. *Studies in Ethnomethodology.* Englewood Cliffs: Prentice-Hall.

Geertz, Clifford. 1973. *Interpretation of Cultures: Selected Essays.* New York: Basic.

Geertz, Clifford. 1983. *Local Knowledge: Further Essays in Interpretive Anthropology.* New York: Basic Books.

Goldmann, Lucien. 1979. "Cinema and Sociology: Thoughts on *Chronique d'un été.*" In *Anthropology—Reality—Cinema,* edited by Nick Eaton, 64–66. London: British Film Institute.

Gorman, Michael. 2010. *Trading Zones and Interactional Expertise.* Cambridge, MA: MIT Press.

Greene, Jeremy A. 2007. *Prescribing by the Numbers.* Baltimore: Johns Hopkins Press.

Grimshaw, Anna, and Amanda Ravetz. 2009. *Observational Cinema: Anthropology, Film, and the Exploration of Social Life.* Bloomington: Indiana University Press.

Hacking, Ian. 1983. *Representing and Intervening: Introductory Topics in the Philosophy of Natural Science.* Cambridge: Cambridge University Press.

Haraway, Donna. 1989. *Primate Visions: Gender, Race, and Nature in the World of Modern Science.* New York: Routledge.

Haffner, Jeanne. 2013. *The View from Above: The Science of Social Space.* Cambridge: MIT Press.

Hawking, Stephen. 1988. *A Brief History of Time.* London: Bantam.

Hocking, Paul, ed. 1974. *Principles of Visual Anthropology.* New York: Mouton de Gruyter.

Johns, Adrien. 1998. *The Nature of the Book: Print and Knowledge in the Making.* Chicago: University of Chicago Press.

Kaiser, David. 2005. *Drawing Theories Apart: The Dispersion of Feynman Diagrams in Postwar Physics.* Chicago: University of Chicago Press.

Kant, Immanuel. 2007. *Critique of Pure Reason.* Translated and edited by Marcus Weigelt. London and NY: Penguin.

Kirby, David. 2011. *Hollywood in Labcoats.* Cambridge, MA: MIT Press.

Kittler, Friedrich. 1999. *Gramophone, Film, Typewriter.* Translated by Geoffrey Winthrop-Young and Michael Wutz. Stanford, CA: Stanford University Press.

Kuhn, Thomas S. (1962) 1970. *The Structure of Scientific Revolutions.* Chicago: University of Chicago Press.

Landecker, Hannah. 2007. *Culturing Life: How Cells Became Technologies.* Cambridge, MA: Harvard University Press.

Latour, Bruno. 1987. *Science in Action: How to Follow Scientists and Engineers through Society.* Cambridge, MA: Harvard University Press.

Latour, Bruno. 1990. "Drawing Things Together." In *Representation in Scientific Practice,* edited by Mike Lynch and Steve Woolgar, 19–68. Cambridge, MA: MIT Press.

Latour, Bruno. 1996. *Aramis, or the Love of Technology.* Cambridge, MA: Harvard University Press.

Law, John and Michael Lynch. "Lists, Field Guides, and the Descriptive Organization of Seeing: Birdwatching as an Exemplary Observational Activity." *Human Studies* 11 (1988): 271–303.

Lezaun, Javier. 2011. "Demo for Democracy." *Limn* 0. Accessed September 9, 2012. http://limn.it/demo-for-democracy/.

Lynch, Michael. 1993. *Scientific Practice and Ordinary Action: Ethnomethodology and Social Studies of Science.* Cambridge: Cambridge University Press.

Lynch, Michael and John Law. 1999. "Pictures, Texts, and Objects: The Literary Language Game of Bird Watching." In *The Science Studies Reader,* edited by Mario Biagioli, 317–342. New York: Routledge.

Lynch, Mike, and Steve Woolgar, eds. 1990. *Representation in Scientific Practice.* Cambridge, MA: MIT Press.

MacDougall, David. 1998. *Transcultural Cinema.* Princeton, NJ: Princeton University Press.

Macfarlane, Allison M., and Rodney C. Ewing. 2006. *Uncertainty Underground: Yucca Mountain and the Nation's High-Level Nuclear Waste.* Cambridge, MA: MIT Press.

MacKenzie, Donald. 1993. *Inventing Accuracy: A Historical Sociology of Nuclear Missile Guidance.* Cambridge, MA: MIT Press.

Marcus, George E. 1995. *Technoscientific Imaginaries.* Chicago: University of Chicago Press.

Masco, Joseph. 2006. *The Nuclear Borderlands: The Manhattan Project in Post–Cold War New Mexico.* Princeton, NJ: Princeton University Press.

Mead, Margaret. 1974. "Visual Anthropology in a Discipline of Words." In *Principles of Visual Anthropology,* edited by Paul Hocking, 3–10. New York: Mouton de Gruyter.

Mialet, Hélène. 2003. "Reading Hawking's Presence: An Interview with a Self-Effacing Man." *Critical Inquiry* 29 (4): 571–598.

Mitman, Gregg. 2009. *Reel Nature: America's Romance with Wildlife on Film.* Seattle: University of Washington Press.

Morris, Errol. 2011. *Believing Is Seeing: Observations on the Mysteries of Photography.* London: Penguin Press.

Pickering, Andrew. 1984. *The Social Construction of Quarks.* Chicago: University of Chicago Press.

Pinney, Christopher. 2004. *"Photos of the Gods": The Printed Image and Political Struggle in India.* London: Reaktion Books.

Rudwick, Martin. 1988. *The Great Devonian Controversy.* Chicago: University of Chicago Press.

Ryle, Gilbert. 1968. "The Thinking of Thoughts: What Is 'Le Penseur' Doing?" *University Lectures* 18. University of Saskatchewan. Accessed August 27, 2012. http://lucy.ukc.ac.uk/CSACSIA/Vol14/Papers/ryle_1.html.

Sahlins, Marshal. 1995. *How "Natives" Think: About Captain Cook, for Example.* Chicago: University of Chicago Press.

Schlosberg, David. 2007. *Defining Environmental Justice.* Oxford: Oxford University Press.

Shapin, Simon, and Steven Schaffer. 1985. *Leviathan and the Air-Pump: Hobbes, Boyle, and the Experimental Life.* Princeton, NJ: Princeton University Press.

Shell, Hanna. 2012. *Hide and Seek: Camouflage and the Media of Reconnaissance.* New York: Zone Books.

Shrader-Frechette, Kristin. 2002. *Environmental Justice: Creating Equality, Reclaiming Democracy.* Oxford: Oxford University Press.

Sibum, Otto. 1995. "Reworking the Mechanical Value of Heat: Instruments of Precision and Gestures of Accuracy in Early Victorian England." *Studies in History and Philosophy of Science* 26: 73–106.

Stoffle, Richard W., and Michael J. Evans. 1992. "American Indians and Nuclear Waste Storage: The Debate at Yucca Mountain." In *Native Americans and*

Public Policy, edited by Fremont J. Lyden and Lyman Howard Legters, 243–265. Pittsburgh: University of Pittsburgh Press.

Tucker, Jennifer. 2005. *Nature Exposed: Photography as Eyewitness in Victorian Science*. Baltimore: Johns Hopkins Press.

Vicedo, Marga. 2009. "The Father of Ethology and the Foster Mother of Ducks: Konrad Lorenz as Expert on Motherhood." *Isis* 100: 263–291.

FILMOGRAPHY

Algar, James. 1951. *Nature's Half Acre*. Walt Disney Productions. 16 mm film, 33 min.

Breton, Stéphane. 2001. *Eux et moi* [*Them and Me*]. Arte Video. VHS, 63 min.

Castaing-Taylor, Lucien, et al. 2009. *Sweetgrass*. Cinema Guild. DVD, 100 min.

Castaing-Taylor, Lucien, and Véréna Paravel. 2012. *Leviathan*. Cinema Guild. DVD, 87 min.

Else, Jon. 1980. *The Day after Trinity*. Pyramid Film and Video. VHS, 90 min.

Flaherty, Robert. 1922. *Nanook of the North: A Story of Life and Love in the Actual Arctic*. Revillon Frères. 16 mm, 65 min.

Galison, Peter, and Pamela Hogan. 2001. *Ultimate Weapon: The H-Bomb Dilemma*. A presentation on the History Channel. GMG Films. DVD, 45 min.

Galison, Peter, and Robb Moss. 2008. *Secrecy*. Redacted Pictures. DVD, 81 min.

MacDougall, David, and Judith MacDougall. 1980. *The Wedding Camels: A Turkana Marriage*. University of California at Berkeley Extension Media Center. VHS, 108 min.

Morris, Errol. 1988. *The Thin Blue Line*. HBO Video. VHS, 101 min.

Morris, Errol. 1991. *A Brief History of Time*. Paramount Pictures. VHS, 80 min.

Moss, Robb. 2005. *The Same River Twice*. New Video. DVD, 77 min.

Negroponte, Michel. 2000. *W.I.S.O.R. (Welding & Inspection Steam Operations Robot)*. New Video. DVD, 75 min.

Negroponte, Michel. 2009. *I'm Dangerous with Love*. Blackbridge Productions in association with Cactus Three. DVD, 85 min.

Oxley, Chris. 2005. *Newton's Dark Secrets*. WGBH Boston Video. DVD, 54 min.

Wiseman, Frederick. 1969. *Hospital*. Zipporah Films. VHS, 86 min.

Wiseman, Frederick. 1974. *Primate*. Zipporah Films. VHS, 105 min.

Wiseman, Frederick. 1987. *Missile*. Zipporah Films. DVD, 115 min.

10 Expanding the Visual Registers of STS

Torben Elgaard Jensen, Anders Kristian Munk, Anders Koed Madsen and Andreas Birkbak

STS is about localization. With that contention Galison sets to work expanding our visual research repertoire. It is thus hardly surprising that he gives pride of place to a practice that is renowned for its ability to capture the local in its richness and depth—namely, that of making films. Indeed, one wonders why the century-old tradition of visual anthropology is only now arousing an appetite among STS scholars that have in other respects been early adopters of ethnographic methods in their pursuit of situated knowledge. Film, Galison argues, offers a "new register in STS" that will complement texts in achieving a "more dimensional, denser understanding of the world of science, technology and medicine." It is hard to disagree.

At least we should be susceptible to the argument since, as Galison writes, we know what kind of essential work the visual can do for science and technology, work that could never be accomplished by texts alone. Why should our research practices be any different in this respect? We have the best of inspirations readily at hand; what remains is to move from first-order to second-order engagements with the visual.

If one takes that idea seriously, however, it becomes interesting to ask if filmmaking is the only way to achieve this. If you consider the multiple roles of the visual in other scientific practices that have been teased out by first-order STS scholarship, filmmaking seems like an unnecessarily narrow banner to be rallying under. We want to suggest that other venerable visual practices from the first-order catalogue might equally deserve to be taken up by second-order visual science and technology studies (VSTS).

In particular, we want to consider the practice of making digital maps, but the argument might be extended to other areas as well. The point is that other-than-filmic visual practices bring other-than-filmic concerns and potentials to the table. VSTS, we would contend, does not have to be exclusively about localizing. It could also be about promoting inquiry or contributing to new assemblages, interventions that might come to the fore if we were to make "visually structured arguments" by means of mapmaking, for example, and not just filmmaking.

THE STRUCTURE OF ARGUMENTATION
AND THE NOTION OF AFFECT

The suggestion to look to digital maps as a new register of social research is not taken out of the blue. Maps have been an important topic in first-order VSTS (e.g., Latour 1986; Turnbull 1996) and critical strands of human geography (e.g., Harley 1989; Cosgrove 1999) for quite a while. Interestingly, though, in parts of the broader social sciences it is increasingly argued that digital maps enable a pure empiricism where honest signals of people's behavior can be studied (Pentland 2008) without the disturbing introduction of theoretical assumptions (Anderson 2008). If digital maps are to function as a register in second-order VSTS it is clearly necessary to conceptualize them differently, taking first-order insights firmly into account, and Galison's arguments are an obvious place to gather inspiration for two reasons.

The first reason is that any visual technology will bring with it a specific theory of knowledge that affords certain modes of argumentation and demands specific skills. Galison illustrates this point nicely in the context of filmmaking. He points, for instance, to the difficulties in transferring oral interview skills to the technology of film, and he shows how the rise of new digital editing tools has altered the ease with which filmmakers can manipulate the pace of the film and engage in nonlinear editing of its story line. Similar points can (and must) be made in the context of digital mapmaking as well.

Take, for instance, the case of harnessing a Twitter stream and turning its digital traces into a semantic map. To produce such a map you need to repurpose the types of data points that Twitter makes available and accept that their structure will guide your engagement with the world. For instance, you need to work with a set of tweets that each have a limit of 140 characters and are structured on the basis of metadata such as hashtags (#) and replies (@). Furthermore, in drawing this data into a map you need to rely on available software for processing large semantic content and comply with its assumptions about natural language (see also Neuhaus and Webmoor 2012).

The point is that the ambition to move a visual register like the digital map from first-order to second-order VSTS needs to be accompanied by discussions of the way it structures argumentation in specific ways. As argued by Galison, we risk losing the specificity of new registers if we fail to adjust our skills to their material affordances and the theory of knowledge they carry. Or—as put by Rogers (2009)—we need to "follow the medium" and its logic when using it as a tool of research. It can therefore never be the position of VSTS to look at digital maps as unbiased mediators of honest signals.

The second reason for turning to Galison's discussion of filmmaking when making sense of digital maps as a register in VSTS concerns his argument that visual technologies leave the viewer *affected* in ways that are different than textual accounts. In the analysis of Wiseman's film about the Yerkes Primate Observatory he, for instance, writes that "having witnessed these scenes . . . no viewer would see this labwork merely as a dismissible path to a crucial conclusion" (Galison 2014, 325). Galison's point is that the film is a visual technology that excels in depicting affect and mixing it with argumentation. There is no disagreeing on our part that film may be a superb medium for generating affective responses, and that maps, in comparison, may appear to be a relatively bland visual register. We would maintain, however, that maps are not irrelevant to affects, but that they simply generate different affects than film.

VSTS BEYOND LOCALIZATION?

If we are to appreciate the potential affectivity of maps, we should probably leave aside notions of affect that tend toward the psychological or physiological; maps rarely shake our nervous systems the way a good film does. The affective relevance of maps changes, however, if we consider some alternative and broader notions of affect, such as the one suggested by Callon and Rabeharisoa (2004). Taking inspiration from both Chinese philosophy and American pragmatism, the authors propose a definition of affect as a network effect: "Pity and compassion are merely the consequences of the entanglement in which we are caught, the attachments constituting us" (2004, 16). Following this they suggest, "This definition opens onto a research programme in its own right, focused on all the mechanisms and material mediations that make these affections, and the entanglements they reveal, visible and perceptible by the actors themselves" (ibid.).

Maps, we suggest, might be one of these "mechanisms and material mediations" that make entanglements visible and perceptible. If an actor is equipped with a map of the multitude of other actors connected to a particular issue, then she is likely to become affected in the sense that new connections and entanglements will become visible to her. This does not necessarily mean that she will become affective in an immediate psychological sense, but it does mean that the maps may visualize her entanglements and that maps may become an occasion for the creation of new entanglements. In our own work with digital maps we have, for instance, experienced how readers of digital maps have achieved a form of reflexivity by seeing themselves and their arguments in relation to other actors and how they have expressed a feeling of curiosity of wanting to explore these other actors in detail.

The possible use of maps and mapping to discover, create or change connections has also been raised by STS scholars (e.g., Marres and Rogers 2005). Taking her point of departure in a Deweyan definition of publics, Marres (2007) argues that a public is the set of actors that are indirectly affected by a particular issue (such as the building of a railroad). In this view, publics are not given; they are sparked into being by issues. From this it follows that material and communicational circumstances are crucial to the formation of publics. To develop new forms of (issue) maps, to trace actors in new ways, and to experiment with new visual communication formats is essentially to contribute to the formation of publics. Mapping, in this view, may be a visual register that enables new forms of entanglement and affectedness.

These publics, of course, are assembled, they are under construction and they are situated and dependent on the devices that have helped bring them into being. But to aid their becoming by visual means is in our opinion to do something distinctly beyond localization. To turn something into the stuff of politics is also to somehow take part in the work of deciding who and what belongs to a certain issue; it is about deciding what should count and how it could be made to count. Democratic inquiry, in the sense of Dewey (1938), has everything to do with becoming aware of what you have at stake in a particular issue, vis-à-vis other stakeholders in that issue, and the ability to navigate those relations.

Attempting to promote such kinds of affectedness, STS scholars are already using visual means to productively slow down collective reasoning in public engagements with science and technology (Whatmore and Landström 2011), to make public the stuff of politics that premises technological democracy (Latour and Weibel 2005) and to provide cartographic devices to aid navigation in messy, controversial or uncertain situations (see Clarke 2003 and Venturini 2012). If we take such examples to be equally plausible candidates for the exciting future of visual scholarship in STS, then perhaps we are not just out to complement and enrich the textual account. We are also out to put things at risk, to energize inquiry, to be more effectively and affectively "idiotic" (in the sense of Stengers 2005) and thus to contribute creatively to the assemblage of new and existing forums.

REFERENCES

Anderson, Chris. 2008. "The End of Theory: The Data Deluge Makes the Scientific Method Obsolete." *Wired Magazine*, July 16.

Callon, Michel, and Vololona Rabeharisoa. 2004. "Gino's Lesson on Humanity: Genetics, Mutual Entanglements and the Sociologist's Role." *Economy & Society* 33 (1): 1–27.

Clarke, Adele E. 2003. "Situational Analyses: Grounded Theory after the Postmodern Turn." *Symbolic Interaction* 26 (4): 553–576.

Cosgrove, Denis. ed. 1999. *Mappings*. London: Reaktion Books.

Dewey, John. 1938. *The Theory of Inquiry*. New York: Holt, Rinehart and Wiston.

Galison, Peter. 2014. "Visual STS" In *Visualization in the Age of Computerization*, edited by Annamaria Carusi, Aud Sissel Hoel and Timothy Webmoor, 308–356. London: Routledge.

Harley, J. Brian. 1989. "Deconstructing the Map." *Cartographica: The International Journal for Geographic Information and Geovisualization* 26 (2): 1–20.

Latour, Bruno. 1986. "Visualization and Cognition." *Knowledge and Society* 6: 1–40.

Latour, Bruno, and Peter Weibel. 2005. *Making Things Public: Atmospheres of Democracy*. Cambridge, MA: MIT Press.

Marres, Noortje. 2007. "The Issues Deserve More Credit Pragmatist Contributions to the Study of Public Involvement in Controversy." *Social Studies of Science* 37 (5): 759–780.

Marres, Noortje, and Richard Rogers. 2005. "Recipe for Tracing the Fate of Issues and Their Publics on the Web." In *Making Things Public: Atmospheres of Democracy*, edited by Bruno Latour and Peter Weibel, 922–936. Cambridge, MA: MIT Press.

Neuhaus, Fabian, and Timothy Webmoor. 2012. "Agile Ethics for Massified Research and Visualization." *Information, Communication & Society* 15 (1): 43–65.

Pentland, Alex S. 2008. *Honest Signals: How They Shape Our World*. Cambridge, MA: MIT Press.

Rogers, Richard. 2009. *The End of the Virtual: Digital Methods*. Amsterdam: Vossiuspers UvA.

Stengers, Isabelle. 2005. "The Cosmopolitical Proposal." In *Making Things Public: Atmospheres of Democracy*, edited by Bruno Latour and Peter Weibel, 994–1003. Cambridge, MA: MIT Press.

Turnbull, D. 1996. "Cartography and Science in Early Modern Europe: Mapping the Construction of Knowledge Spaces." *Imago Mundi* 48 (1): 5–24.

Venturini, Tomasso. 2012. "Building on Faults: How to Represent Controversies with Digital Methods." *Public Understanding of Science* 21 (7): 796–812.

Whatmore, Sarah J., and Catharina Landström. 2011. "Flood Apprentices: An Exercise in Making Things Public." *Economy and Society* 40 (4): 582–610.

11 Mapping Networks
Learning From the Epistemology of the "Natives"

Albena Yaneva

As an anthropologist trained by Bruno Latour, my research has taken me to fieldwork sites as exciting as the *Musée d'art modern de la ville de Paris* and the Office of Metropolitan Architecture in Rotterdam in 2001, among others. At the time of these studies, Actor-Network-Theory (ANT)[1] was taken beyond its privileged domains of action (science, technology and engineering as praxis) and was now being used as a method to investigate other fields spanning architectural design, the arts and markets. Akin to the scientific laboratory as rediscovered by STS scholars in the 1980s, the daily routines of image production by artists, technicians, curators, architects and urban planners I followed since the 1990s provided me with a fertile source of predominantly visual data.

I've learned that the ethnographic images do not merely document a practice in minute detail but may themselves shape explanations, strengthening or diminishing arguments. In going beyond illustration, they mediate, they travel in cascade (Lynch 1985; Latour 1990). Their manner of exploration thus captures the density of local "cultures" of practice while questioning what matters to practitioners: what truly defines their practices. Over many years of thick empirical exploration I became increasingly interested in the techniques of image production used by the aforementioned "tribes" of designers and artists. As their work itself (like that of scientists) relies so heavily on visuals, my studies led me not only to analyze a multitude of images but also to engage myself in the drawing of maps and charts, in the cutting of foam for scale models. These studies were themselves largely inspired by the first-order visual STS (or VSTS) coined by Peter Galison.

ANT defines knowledge practices as restricted and circumscribed into intricate and frangible, yet costly, networks of praxis. In attempting to visualize and trace such networks, a number of collaborations between the fields of STS and web technologies were initiated (Callon 2001). At the same time, there was much debate within STS regarding new ways of shifting the ANT focus beyond single-sited ethnographies toward methods able to grasp the "figurational" dimensions of collaborative networks. Rather than simply replacing old methodologies with the current sociological alternative of the time (which would reduce "figurational" complexity to a few quantitative indicators and so destroy the specificity of the phenomena under scrutiny) STS scholars opted for a combination of methods. As well as interviews and content analysis, this involved the use of a software program, *Réseau Lu*, specifically designed to map complex, heterogeneous, relational data in order to render it visually interpretable (Cambrosio, Keating and Mogoutov 2004; Cambrosio et al. 2010). A second web technique, Issue Crawler, allowed researchers to explore online debates by tracking the relationships between the actors involved via hyperlinking as well as mapping of their discursive affinities (Rogers 2004). Thus, we had witnessed a shift toward the second order of VSTS, whereby digital media penetrated the humanities, concomitant with the entrance of parametric design and BIM[2] into architectural design.

The underlying assumption of these scholars was that hyperlinking and discursive maps provide a semblance of these web-based socio-epistemic networks (Marres and Rogers 2000), thus enabling scholars to witness, in the process of mapping, the making of a debate whose format and magnitude might not be initially obvious. These were, at the time, pioneering attempts to combine qualitative and semiquantitative methods. Notably, rather than replacing ethnographic accounts or substituting the painstaking collection of ethnographic data itself, these maps were intended to inform, enrich and complete an enquiry conducted with qualitative methods. Maps were neither an end point nor standby for visual (qualitative or quantitative) methods. They instead provided a starting point for a richer and more intense enquiry and thus greatly facilitated the generation of a novel type of qualitative data.

The use of Issue Crawler and *Réseau Lu* maps constituted a major step forward in tracing these networks visually and analytically, thus enabling STS to powerfully respond to critiques of its predominantly qualitative methods. Networks cannot be reduced to pure social relations: They are *socio-technical*, and the power of the mapping techniques lies in their capacity to trace the heterogeneous constitution and dynamics of these networks. To take an example from my own fieldwork, it is clear that two years in an architectural practice cannot simply be replaced with a figurational map. However, an understanding of such a practice could be much enhanced, strengthened and complemented by a mapping technique that adequately situates it within a larger comprehensive network. Made up of the many agencies that make it work, this network map would in turn allow for the inspection of the actors' constitutive relationships. Such a rendering could provide a visual strength to the analytical arguments made on the basis of ethnography only, and would allow for a more comprehensive view of the larger-scale networks of architectural production.

Though I never produced maps at the time of my ethnographic study of the museum in Paris and the architectural practice in Rotterdam, I continued to travel and do fieldwork. In studying the practices and beliefs of different tribes of artists, architects, designers and craftsmen, their strange obsession with time, novelty and innovation, and their enigmatic attachments to sketches, drawings and maps, I also started to learn a great deal from their indigenous visual epistemologies. Reflecting further on the type of *visual* knowledge that I could take back to STS and ANT immersed me in a dialogue with architects and designers as part of the controversy mapping collaborative project MACOSPOL, facilitated by Bruno Latour (2008–2010, EU grant). The following paragraphs will present some of the subsequent developments on the aforementioned network mapping, alongside reflections on the type of VSTS I have concurrently been able to develop.

In the academic context of their creation (the introduction of the second-order VSTS) the network maps developed through the collaborations of STS scholars and web specialists were highly productive and innovative. Yet returning to them now, a decade later, we feel it is necessary to refine and improve upon two aspects of the network maps in particular. Firstly, though the web browsers they are based upon are by definition dynamic, the final maps are static and difficult to manipulate and update. They do not satisfactorily convey the "processual" aspect of phenomena explored by ANT. Secondly, their visual repertoire draws on rather limited graphical conventions. With nodes standing for actors and lines for connections, we had produced what is in a technological and visual sense effectively a chart made up of nodes at the conjunction of two intersecting lines. The resulting map is thus "anemic" in its representation, in the words of Peter Sloterdijk (in Latour 2009), who drew our attention to the distinction between networks and spheres. In order to address the limitations of the existing network visualizations and improve their visual power we recognized the need to overcome this "anemic" aspect.

In the intense dialogue with architects and computational designers that followed I tried—rather than studying their practices with STS toolkit in hand—to learn from them a different mode of "doing" STS, one that was more dynamic and graphically innovative. Though unaware of this at the time, I was effectively "doing" VSTS. The computational designers and architects became my anthropologists, while ANT-ers (myself and colleagues) became their "tribe." Identifying the visual techniques that natives use to generate reality before transporting them into a field of study where these techniques can further knowledge about native phenomena might appear as an unconventional technique. Ought we to use a master plan to study the practices of planning? Statistic graphs to study statisticians at work? Scale models to study architects and collages to study artists? There is something that fascinates me in that extreme way of *learning* from the epistemology of the "tribes" we follow, *how* to study them in a different way. That is an intriguing shift in epistemology.

This reverse anthropology allowed me to learn what computational design can offer, while they in turn learned what ANT stands for and provided answers to our many questions. Our aims were twofold. The first was to develop a map in which the actors' identities would not be determined prior to the mapping and *outside* of it, fixed in color dots and bubbles. We wanted to observe the process through which the actors gained identity via tracing their dynamic interconnection and express this in a different visual language. The second aim consisted in overcoming the *static* representations of networks. Ideally, we envisioned a mapping able to provide a dynamic and user-friendly interface, in order to capture the constituency of a network and the fluency of the social. Such a mapping would have the potential to make stronger epistemological statements, while overcoming the ontological assumption of "out-there-ness."

Circumventing the allegiance to rigid geometrical figures, computational design (and parametric design in particular) offers flexible ways of visualizing the dynamic relations of all technical features that generate a new shape. It depicts a much more fluid world of dynamic entities in which the form of the phenomenon does not appear as determined from outside; it is *being* shaped in the geometric flows. A simulation relies on the assumption of subjects and objects that cannot be isolated from their surroundings but appear in continuity with them; it points to a "mutually shaping process through which the dynamic of formation is supposed to concern simultaneously subject and object" (Rahim 2006, 136). This specific epistemology of computation helped us to develop an analytical method capable of portraying a degree of complexity that is difficult to capture with traditional qualitative and quantitative methods.

Through engaging in an active dialogue with computational designers, computation itself transmogrified into something more than a site of ANT-inspired exploration. It became a site of mutual experimentation of observers and natives. The roles constantly shifted and professional boundaries blurred. This allowed us to visualize how an assembly of heterogeneous actors, their locations in time and space, and conflicting concerns work in tandem to shape the controversy. It also helped us to witness the particular fluency inherent in, for example, the making of the 2012 London Olympics stadium, which is typically treated as a static entity (Figure 11.1). The mapping captures the fluidity of the social and thus better accounts for its temporal features. As we follow the simulation, we also witness how each element in the mapping constitutes the actual, with the ongoing rendered as an original element in the constitution of other actual entities, actors and concerns that are deployed and elicited by repetitions of process. Far from representing existing social features, the dynamic network mapping captures occurrences, events and situations that make the social traceable, graspable.

Can this mapping replace or complement a conventional ethnography of the process of design and construction of the 2012 London Olympics stadium? Is it infra- or metareflexive? Is it "spheric" enough, or is it still "anemic"? What kind of tool is it? Mapping has an important performative force. Instead of predefining the content and format of the particular debate, one can witness (as the mapping unfolds in the simulation) the making of actors, alliances and concerns that might not be obvious at the start.

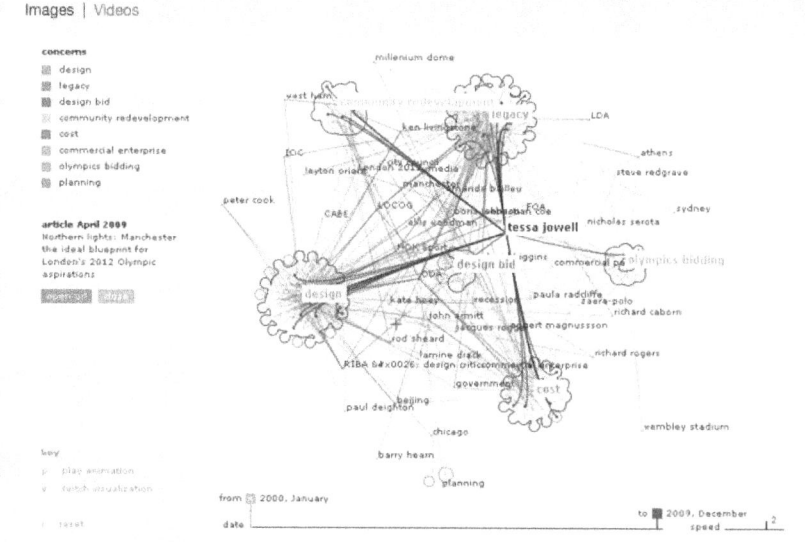

Figure 11.1   The dynamic network mapping of the process of design and construction of the 2012 London Olympics Stadium. http://www.mappingcontroversies.co.uk/london/

Mapping is an epistemological tool for underlining ontological singularities: a powerful visual device for deploying, not just describing, phenomena. Indeed, this provides us with a technique for producing infrareflexive accounts of socio-technical objects and processes. Mapping is not a way of illustrating, but a way of generating and deploying knowledge. It has its own range of epistemology, offering a denser understanding of the world, in addition to textual descriptions and analyses.

In this experiment of swapping tools with the natives, and shifting methodologies, I experienced a double challenge. As well as striving to mobilize architectural tools in producing an improved representation of the hybrid social phenomenon of stadium making, we were simultaneously learning from the epistemology of computational designers a better way of conducting ANT-inspired research. If the first-order VSTS used images as its subject matter of inquiry, second-order VSTS in turn used visuals as a form of generating knowledge about the practice itself in leading to the visual structuration of arguments. A further way of pushing such visual structuration forward, I suggested here, is to borrow epistemologies and methodological insights from the "natives," which leads to an intriguing way of shifting interpretative frameworks in taking positions that will not be always as convenient as the one embraced by the participant observer.

This method of dynamic mapping and simulation of controversies is new to the field of controversy studies in STS, as well as to architecture studies (Yaneva 2012). It succeeds in tracking controversies in motion while also innovating the graphical repertoire of network visualizations, thus offering a new take on complex social phenomena.

NOTES

1. ANT is an approach of "sociology of associations" and a method of enquiry developed by Bruno Latour, Michel Callon, and further by John Law. It is an anti-essentialist way of interpretation and description of networks by "following the actors".
2. Building Information Modeling (BIM) is a process involving the generation and management of digital representations of physical and functional characteristics of places. BIMs are files, which can be exchanged or networked to support decision-making about a place.

REFERENCES

Callon, Michel. 2001. "Les méthodes d'analyse des grands nombres peuvent-elles contribuer à l'enrichissement de la sociologie du travail?," In *Sociologies du travail: Quarante ansaprès*, edited by Amélie Pouchet, 335–354. Paris: Elsevier,

Cambrosio, Alberto, Pascal Cottereau, Stefan Popowycz, Andrei Mogoutov and Tania Vichnevskaia. 2010. "Analysis of Heterogeneous Networks: The RéseauLu Project." In *Digital Cognitive Technologies*, edited by Claire Brossard and Barnard Reber, 137–152. London: Wiley.

Cambrosio, Alberto, Peter Keating and Andrei Mogoutov. 2004. "Mapping Collaborative Work and Innovation in Biomedicine: A Computer Assisted Analysis of Antibody Reagent Workshops." *Social Studies of Science* 34 (3): 325–364.

Latour, Bruno. 1990. "Drawing Things Together." In *Representation in Scientific Practice*, edited by Michael Lynch and Steve Woolgar, 19–68. Cambridge, MA: MIT Press.

Latour, Bruno. 2009. "Spheres and Networks: Two Ways to Reinterpret Globalization." *Harvard Design Magazine* 30 (spring/summer): 138–144.

Lynch, Michael. 1985. "Discipline and the Material Form of Image: An Analysis of Scientific Visibility." *Social Studies of Science* 15: 37–66.

Marres, Noortje, and Richard Rogers. 2000. "Landscaping Climate Change: A Mapping Technique for Understanding Science and Technology Debates on the World Wide Web." *Public Understanding of Science* 9 (2): 141–163.

Rahim, Ali. 2006. *Catalytic Formations: Architecture and Digital Design*. London: Taylor & Francis.

Rogers, Richard. 2004. *Information Politics on the Web*. Cambridge, MA: MIT Press.

Yaneva, Albena. 2012. *Mapping Controversies in Architecture*. Farnham: Ashgate.

12 If Visual STS Is the Answer, What Is the Question?[1]

Anne Beaulieu

Calls to reflect on turns, such as a visual turn in STS, are opportunities to point to all the wrong reasons for making such claims. "We have always been visual," say the critics. There is "no need to jump on the infographics bandwagon," nor to pander to new, larger audiences by turning to video, film festivals and YouTube. "Why allow yourself to be seduced by the wow effect of visuals?" ask others. Even those who argue that they are simply following the actors and entering their visualizationized worlds could be accused of going native, of fetishizing the visual.

Across all these critiques of a turn to visual STS is the implicit construction of the visual as *other* to STS. One way to set up this otherness is to oppose it to what the STS scholar possesses, so that the visual is positioned as what *others* have (the NASA scientists, the users of big scanners, the simulation-builders, the artistically inclined scientific geniuses). Another version is to position the visual as what we as STS scholars bring back from the field, enabling us to blind our audiences with (visual) science. Yet another move is to position the visual as what STS doesn't yet have and which must be introduced, at the risk of disturbing conventions and expectations—perhaps even forbidding colleagues from taking photos during a presentation, as happened at the Visualization in the Age of Computerization event, because visual material falls under a different regime of intellectual property.

The othering of the visual achieves what othering (of nature, women, etc.) in other instances has done, and as an intellectual move, it can be seen as just another case of boundary-drawing. The question of visual STS does become interesting in its specifics. A potential prophylactic to the fetishization of the visual may therefore be to focus on the "problematization," following the work of scholars such as Annemarie Mol and Paul Rabinow. Rather than ask in general terms about what is the visual term (and what is its other), my proposal is to ask: How/when/where/for whom does visualization become worth examining and analyzing? This seems especially important to consider carefully, given that Burri and Dumit's (2008) canonical review documents ample presence of visual objects in STS, while analysis of the use of the visual for scientific discovery and reasoning (Gooding

1990, 2003; Rudwick 1976) has also been present since the early days of STS. As a method and approach for STS, notably, the use of photography can be traced back to *Laboratory Life*, and video ethnography was alive and well over a decade ago at the 4S conference in Atlanta in the hands of Richard Duque, Wesley Shrum and colleagues. The visual, it would seem, has already been found to be the answer to a range of fascinating questions. In the rest of this piece, I put forth a number of key themes that might be valuable as concepts in articulating the practice of visual STS.

EMERGENT

Some of the most interesting aspects of the visual may be discovered by following Christine Hine's expectation that "qualities of digital and physical objects will be emergent in contexts of practice" (Hine 2012, 3). What the visual can and cannot do then becomes less of an a priori insight, and forms part of the problem of the visual in science and in STS. Furthermore, the visual can be found in forms that place it in many different categories—as surfaces (graphic on a screen), tools (visually enhanced instruments), interfaces (data visualization), things (printed x-ray), texts (when showing traces and measurements) and even finding a place within a cultural logic of puzzles, intrigues and revelations (Saunders 2009).

The variations in the visual not only tell us about its rich material culture but also lead us to consider the differentiated expectations, labor, skill and experience involved in encountering visuals. They also tell us a lot about the wide variation in the ontology of images (Carusi, Hoel and Webmoor 2012; de Rijcke 2013), opening up their analysis to possibilities beyond those of representation and visual perception.

BORING

Would a practice of visual STS be more concerned with aesthetics than the usual STS? If the visual has been associated with the exciting or seductive (Stafford 1991), studies of visualizations can actually reveal that working with images is just as boring as dealing with any other set of scientific traces, and that their handling is no more glamorous than any other set procedures pursued in infrastructures. Indeed, one of the common sites of much "visual" work in the age of computerization is the operator behind the console and keyboard, who declares to the investigator (like, indeed, so many of our objects of study do) that they are actually not doing anything worth observing. "Behind the screen" is a difficult site, where the space of the screen and the space of work weave together in an interface whose complexity and distributedness may be challenging to apprehend (de Rijcke and Beaulieu 2011). Following screen-work requires determination, saintly patience and methodological adaptation. Video will record the movements of the operator, and screen grabs or even screen videos might capture other aspects, but screen-work and computation, at the heart of so many visualizations, are neither exciting nor easy to observe. And like in much of STS, while the outcomes may be gorgeous and exciting (though there are exceptions—see Coopmans 2011), visual STS may very well find that it partakes in the STS tradition of the study of boring things (Star 1999).

OBSERVER

The very word visual relies on a perceptual category, that of sight. The visual is that which is seen. But who is doing the seeing? This is potentially one of the most fascinating ways in which the visual in the age of computerization is being problematized. There is a range of ways of talking about seeing in the practice of visual STS. Across the chapters in this volume, we encounter viewers, users, observers, trainees, agents, scientists, visitors (to websites), technicians, doctors, etc. Among these, there are a range of observers—humans and machine, alive and dead (Crary 1990), trained and naïve, mathematically and biologically driven, mechanically consistent or "learning." Given this range of identities, there are a correspondingly broad range of ways of learning to see, whether it be learning to see like an expert (Goodwin 1995) or like a colleague or scholar (Carusi 2011) or whether it involves issues of calibration and pipelines (de Rijcke and Beaulieu 2014). While automated optical recognition remains an AI dream (albeit a powerful one), the visual is a terrain where the social and technological emerge (Mayer 2012), and where the visual as both distributed cognition and embodied knowledge is especially intriguing. In a practice of visual STS, the conventional understanding of the visual as relating to sight is shown to be an achievement rather than a starting point.

NETWORK

If there are multiple forms of observing, it is not surprising that there are a variety of forms of witnessing (Woolgar and Coopmans 2006), most recently across algorithms and interfaces. The idea of the visual as more grounded in place and time is part of the opticist regime, where a naturalistic view of seeing posits that the human eye is the privileged recipient of the light particle that has touched the object and then travels to the retina. This view is often extended to many media through assumptions of their transparent mediation[2] that does not breach the privileged chain of connection inherent in the opticist regime.

Recent developments around the use of networks and information visualization, among other interfaces, mean that the visual is involved in witnessing in forms that are not primarily optical. Authoritativeness and truth effects of these images are achieved through particular kinds of circulation and mediation in networks, a phenomenon de Rijcke and I have termed "network realism" (Beaulieu and de Rijcke, forthcoming). On-the-fly generation, calculation and interaction, rather than a physical theory of the transmission of photons, constitute the conventions that produce the visual as real. Databases, networks, analysis pipelines and interfaces become increasingly important for the visual, and have growing purchase over notions of looking, surface and representation. Circulation rather than capture may be the main mode of contemporary visual culture.

TOOLKIT

Luckily for STS, visual or plain old, there are several areas of work that provide tremendously useful resources for STS work concerned with the visual.[3] In media studies, film studies and art history, the visual has been in question, as have modes of mediation (Bolter and Grusin 1999) and issues of materiality in relation to production, viewing and circulation of images (Wasson 2005; Acland and Wasson 2011). Feminist critiques of visuality (including incursions into STS via the work of Donna Haraway and Sharon Traweek, among others) are also especially useful exemplars to tease out the political and ethical dimensions of the visual and powerful antidotes to visual romanticism and its holistic claims.

If there is a question to be posed in STS, it is best pursued in the form of projects

- that see images as particular points of interventions and emergence
- that seek to understand how the network generates its own form of realism
- that aim to follow hybrid framings that result from combinations of analysis pipeline and expert users
- and that care to look at both screen-work and screen pleasure.

Such explorations are likely to be the most satisfying answers to the question of visual STS.

NOTES

1. While the framing for this reflection was provided by the editors of this volume, the arguments it contains were put forth in my closing remarks at the Visualization in the Age of Computerization conference in Oxford in March 2011 and further refined in the course of the discussion that followed.
2. See Bolter and Grusin (2000) on mediation (especially through digital media) and Ashmore, MacMillan and Brown (2004) on the fetishization of auditory traces.

3. See the recent 4S/EASST panel in Copenhagen in 2012, Mediated Practice: Insights from STS, Critical Theory and Media Theory (http://www.audsisselhoel.com/wordpress/?p=133). Last accessed 4 November 2013.

REFERENCES

Acland, Charles R., and Haidee Wasson, eds. 2011. *Useful Cinema*. Durham, NC: Duke University Press.

Ashmore, Malcolm, Katie MacMillan and Steven D. Brown. 2004. "It's a Scream: Professional Hearing and Tape Fetishism." *Journal of Pragmatics* 36 (2): 349–374.

Bolter, Jay D., and Richard Grusin. 2000. *Remediations: Understanding New Media*. Cambridge, MA: MIT Press.

Beaulieu, Anne, and Sarah de Rijcke. Forthcoming. "Networked Knowledge and Epistemic Authority in the Development of Virtual Museums." In *Museum Transfigurations: Curation and Co-creation of Collections in the Digital Age*. Edited by Chiel van den Akker and Susan Legêne. New York: Berghahn Books.

Burri, Regula V., and Joseph Dumit. 2008. "Social Studies of Scientific Imaging and Visualization." In *The Handbook of Science and Technology Studies*, 3rd ed., edited by Edward J. Hackett, Olga Amsterdamska, Michael E. Lynch and Judy Wajcman, 297–318. Cambridge, MA: MIT Press.

Carusi, Annamaria. 2011. "Computational Biology and the Limits of Shared Vision." *Perspectives on Science* 19 (3): 300–336.

Carusi, Annamaria, Hoel, Aud Sissel and Timothy Webmoor. 2012. "Editorial." In "Computational Picturing." Special issue. *Interdisciplinary Science Reviews* 37 (1): 1–3.

Coopmans, Catelijne. 2011. "'Face Value': New Medical Imaging Software in Commercial View." *Social Studies of Science* 41 (2): 155–176.

Crary, Jonathan. 1990. *Techniques of the Observer: On Vision and Modernity in the Nineteenth Century*. Cambridge, MA: MIT Press.

de Rijcke, Sarah. 2013. "Staging the Studio: Enacting Artful Realities through Digital Photography." In *Hiding Making—Showing Creation: The Studio from Turner to Tacita Dean*, edited by Rachel Esner, Sandra Kisters and Ann-Sophie Lehmann, 226–244. Amsterdam: Amsterdam University Press.

de Rijcke, Sarah, and Anne Beaulieu. 2011. "Image as Interface: Consequences for Users of Museum Knowledge." *Library Trends* 59 (4): 664–685.

de Rijcke, Sarah, and Anne Beaulieu. 2014. "Networked Neuroscience: Brain Scans and Visual Knowing at the Intersection of Atlases and Databases." In *Representation in Scientific Practice Revisited*, edited by Catelijne Coopmans, Janet Vertesi, Michael Lynch and Steve Woolgar, 131–152. Cambridge, MA: MIT Press.

Gooding, David C. 1990. *Experiment and the Making of Meaning: Human Agency in Scientific Observation and Experiment*. Dordrecht: Kluwer Academic.

Gooding, David C. 2003. "Varying the Cognitive Span: Experimentation, Visualization and Computation." In *The Philosophy of Scientific Experimentation*, edited by Hans Radder, 255–283. Pittsburgh: University of Pittsburgh Press.

Goodwin, Charles. 1995. "Seeing in Depth." *Social Studies of Science* 25: 237–274.

Hine, Christine. 2012. "The Emergent Qualities of Digital Specimen Images in Biology." *Information, Communication and Society* 16 (7): 1157–1175.

Mayer, Katja. 2012. "Objectifying Social Structures: Network Visualization as Means of Social Optimization." *Theory & Psychology* 22 (2): 162–178.

Mol, Annemaire. 2002. *The Body Multiple: Ontology in Medical Practice*. Durham, NC: Duke University Press.

Rabinow, Paul. 2003. *Anthropos Today: Reflections on Modern Equipment.* Princeton, NJ: Princeton University Press.

Rudwick, Martin J. S. 1976. "The Emergence of a Visual Language for Geological Science 1760–1840." *History of Science* 14: 149–195.

Saunders, Barry F. 2009. *CT Suite: The Work of Diagnosis in the Age of Noninvasive Cutting.* Durham, NC: Duke University Press.

Stafford, Barbara Maria. 1991. *Body Criticism: Imaging the Unseen in Enlightenment Art and Medicine.* Cambridge, MA: MIT Press.

Star, Susan L. 1999. "The Ethnography of Infrastructure." *American Behavioral Scientist* 43 (3): 377–391.

Traweek, Sharon. 1999. "Warning Signs: Acting on Images." In *Revisioning Women, Health, and Healing: Feminist, Cultural, and Technoscience Perspectives*, edited by Adele Clarke and Virginia Olesen, 187–201. New York: Routledge.

Wasson, Haidee. 2005. *Museum Movies: The Museum of Modern Art and the Birth of Art Cinema.* Berkeley: University of California Press.

Woolgar, Steve, and Catelijne Coopmans. 2006. "Virtual Witnessing in a Virtual Age: A Prospectus for Social Studies of E-science." In *New Infrastructures for Knowledge Production*, edited by Christine Hine, 1–25. London: Idea Group.

13 Visual Science Studies
Always Already Materialist

Lisa Cartwright

In 1987 *Feminist Studies* published "Fetal Images: The Power of Visual Culture in the Politics of Reproduction," an essay about obstetrical ultrasound by the political scientist Rosalind Pollack Petchesky. Considering the boundaries between media spectacle and clinical experience, Petchesky's essay catalyzed debates about gender and sexuality in the visual culture of science and medicine, doing for science, technology and medicine studies[1] what Laura Mulvey's "Visual Pleasure and Narrative Cinema" (1975) had done for film and media studies more than a decade earlier. Like Mulvey, Petchesky addressed the issue of pleasure and its connection to the nexus of subjectivity, sexuality and power in the analysis of experiences organized around an imaging technology, a social practice, and human subjects situated together in a given field of practice.[2] A year after Petchesky's essay appeared, *Human Values* released a special issue about images and imaging technologies in scientific practice. Drawing variously from work in the sociology of laboratory studies, ethnomethodology of science, and the semiotics of scientific iconography, the editors Michael Lynch and Steve Woolgar republished this collection as *Representation in Scientific Practice* (1990), a book which would become a classic in the emerging canon devoted to visuality and images in science studies. A point often missed in the historiography of the field is that the Marxist materialist focus on technology and practice by scholars associated with the sociology of scientific knowledge (SSK), a locus of research that emerged in the 1970s prior to the visual turn in science studies, was congruent with a film studies focus on apparatus, embodiment and the means of production during that same period. Science studies and film studies scholars shared an interest in the Althusserian Marxist approach to the apparatus and means of production; however, whereas scholars in film studies foregrounded Althusser's interest in the Lacanian theory of the human subject and intrapsychic experience, this focus on subjectivity was less prevalent in science studies work on the apparatus. Petchesky's essay was one of the few exceptions, and might in fact be regarded as tangential to the science studies field insofar as it is rarely remembered and taught in the field. Medical and scientific illustrations and photographs had become a topic of interest among historians of art, science and medicine in the 1980s, with Linda Dalrymple Henderson producing an important book on the science-art nexus (Henderson 1983), and Ludmilla Jordanova publishing her now classic book about scientific representations of gender and sexuality in medicine (Jordanova 1989). But those contributions did not include insights from the film and visual studies focus on the apparatus and the place of the body in the material means of image production. In 1992, the journal *Representations* devoted an issue to the topic of "Seeing Science," including an essay by historians of science Lorraine Daston and Peter Galison that became a classic in the science studies literature for its attention to the place of images in scientific objectivity (Daston and Galison 1992). That year also saw the publication of *Techniques of the Observer*, a book by art historian Jonathan Crary about nineteenth-century vision and modernity that became one of the most widely cited non–science studies works on visuality in the field, and one of the few works offering substantial consideration of optics in the history of philosophy. But those contributions, like the SSK literature, did not centrally feature subjectivity apart from embodied optics and knowledge in the figure of the scientific observer.

Whereas Petchesky's focus was on subjectivity, pleasure and the embodied, affective experiences of patients or clients along with medical scientists and clinicians, the essays in Lynch and Woolgar's collection and the article by Daston and Galison foregrounded objectivity, knowledge and the professional practice of researchers and scientists. In 1991, the feminist film studies journal *Camera Obscura* brought out the first of two special issues on the topic of imaging technologies in science and medicine, introducing the matter of gender, sexuality and representation in science and medicine through works by feminist scholars, artists and activists in linguistics, medical anthropology, medical humanities, film and media studies, and AIDS and cancer journalism and activism.[3] Like Petchesky's essay, the articles in these two issues emphasized the experience of research subjects, patients and clients. The embodied practice of scientists and doctors using images and imaging technologies was discussed, though primarily from a relational standpoint in which the experience of human subjects acted upon in the research or clinical context was the central concern.

I offer this quick sketch of items from the literature on visuality and representation in science studies to suggest that we might speak of a visual turn in the field of science studies around 1990, a turn that entailed a range of emphases along the spectrum of objectivity and subjectivity. Whereas objectivity and knowledge would emerge as dominant concerns in science studies work about visuality, subjectivity, situated knowledge and the phenomenology of experience were consistently a stronger presence in both feminist science studies and the sociology and anthropology of medicine. Feminist epistemology of science, though focused on knowledge, drew considerable attention to the matter of embodied standpoint and to subjugated knowledge. Writing in phenomenology, ontology and historical materialism brought forward bodies—those of research subjects and patients as well as laboratory workers and scientists or clinicians—as crucial components of science studies research, and as features of the material apparatus.[4]

This brief and speculative signposting of directions in the field sets the ground for a series of proposals about the place of the senses and feminist materialism in the recent return to discussions about the visual in science studies. At this historical moment, one could say that science studies has seen both a sensory and a materialist turn. In this context of the mid-2010s, I wish to gesture to an important set of feminist precedents in the science studies field—work in visual science studies that may be understood to constitute a precedent in the regard of the visual, and to have initiated an ongoing set of methods, theories and concerns organized around subjectivity and embodiment and incorporating materialist feminist methodologies in the decades that bridged the centuries.

My first proposal is that science studies as a field historically has for the most part adopted approaches from visual studies that minimize subjectivity in favor of attention to objectivity. Work on subjectivity has been heavily concentrated in approaches to embodiment and materialism in ways that account for the professional practice of the researcher, will less attention to the experiences of human research subjects or patients. The legacy of feminist materialist work on representation and embodiment, though present in the feminist anthropology, sociology and history of medicine (Martin 1987, Clarke and Moore 1995, Herzig 2005), has been underutilized in its potential to bring balance across this spectrum. Determining to what degree and how visual studies and science studies were interarticulated fields throughout their development is a question worth posing in order to understand the relatively stronger emphasis on objectivity. The visual turn of the 1980s and 1990s prompted visual studies to emerge, like science studies before it, as a field in its own right. During the decades that spanned the centuries, visual studies gained a journal, university programs and professional societies. Yet science studies has remained marginal to the visual studies canon, with only a few scholars explicitly claiming participation in both established disciplines.[5] And visual studies has informed science studies in highly selective ways, mostly through the concept of scientific vision and objectivity. Some science studies scholars whose work has been significant to discussions about images and the visual have suggested that images are a different matter in science than in art or everyday culture. For example, in a brief contribution to *Representation in Scientific Practice Revisited* (Coopmans et al. 2014), Bruno Latour remarks that "focusing on the visual per se might lead in the end to a blind alley. The reason is that image making in science is very peculiar." Ahead I consider this attitude of visual exceptionalism as integral to visual studies as a field, suggesting that the specificity of visual practice in fact should lead precisely to work on the visual. Readers may note that the notion of a "blind alley" attributes the condition of "not being able to see" not to the sensory body but to the space itself—it is the alley that is deemed "blind" and not the human subject who passes through it. This locution places the emphasis on the field of the gaze—precisely the point Mulvey made in her classic essay on sexuality and representation, and precisely the issue at stake in Petchesky's essay on fetal images. For Petchesky, the interior, invisible space of the womb—its status as, one might say, a "blind alley"—is rendered a field of the gaze precisely because of the matter of what can and cannot be sensed. She highlights the function of the fetal sonogram, a technology of sound, in its paradoxical appearance as an image, an entity that signifies a subject and a being that has yet to take full material form. The impasse of visuality that is physical interiority and minuteness becomes an occasion for staging presence. Rather than turning away from a visual impasse, the "blind alleys" that confuse and disappoint the hunt for visual evidence, we might instead interrogate the material terms of this space in order to understand the relationship between the image and that to which it obliquely gestures.

My second proposition is that feminist research about images, representation and visuality has been critical to the development of the interpretation of images and imaging in science, technology and medicine studies. The feminist work in question has held close to a materialist philosophy, grounding the study of scientific visuality in a phenomenological approach to embodied experience. The matter of how one negotiates the impasse to which Latour refers is precisely what is at stake in this earlier feminist work as well as in some of the writing in the current visual turn. I will be proposing that the current attention to the visual in science studies is in fact in the context of a sensory turn in which scholars have been cognizant of precisely the "peculiar" manner in which the visual tends to mislead, or to lead elsewhere, to draw attention to other places than the location of image or of referent, and to lead us to see complexity and the distributed nature of situated visual practice, as in the paradoxical case of the sound image. The sensory turn in science studies, I will propose, moves away from a concern with objectivity and epistemology and toward theories of subjectivity, ontology and matter. Ahead I will be discussing visual science studies research in the sensory turn and this focus on subjectivity, ontology and matter through an extended discussion of the work of sociologist and videographer Christina Lammer.

My third proposition is that we take note of the situation of the visual turn in science studies in the context of the digital turn. Visual science studies of the 1990s coincided with the mass introduction of digital technologies, including digital imaging technologies, in science, clinical medicine and academic publishing.[6] Digital imaging systems linked to technologies such as CT, MRI and PET transformed radiology from an image-based field into a domain of quantitative visual data. In the 1980s and 1990s MRI, ultrasound and new digital techniques in CT and conventional radiological practices transformed biomedical research fields and clinical medicine. By 1990, "image" had begun to serve as an anachronistic reference point to older techniques and practices as the interpretive focus in biomedical scientific practice shifted to networks, systems, signals and information. The coincidence of the visual turn with the digital turn was made evident in the new media forms used in biomedical research and clinical practice in the 1980s. Likewise, the biotechnology imaging industry coincided with transformations in the global economy, impacting science and medicine, but also impacting science studies research methods and publishing formats. With digital reproduction and with outsourced manufacture lowering production costs in publishing, images could be reproduced in abundance, color images could be incorporated with ease into science journals, and science studies scholars could access and write about these techniques with relative ease. Theory and method in interdisciplinary science studies were changed not only by the observational accessibility of the visual methods used in contemporary biomedicine but also by the digital turn's introduction of techniques for conducting research in the sociology, philosophy and history of science and medicine. A flurry of academic and lay press accounts documenting these new imaging technologies in medicine and science coincided with attention by science studies scholars to images and imaging practices among physicians and scientists engaged in the digital turn that coincided with the visual turn of the 1980s. As the field of radiological imaging was transformed and ultrasound and MRI were brought into routine diagnostic, clinical and surgical practices, science studies researchers took digital video recorders, still cameras and other sorts of tools into the clinic and the laboratory. The digital turn, along with the outsourcing of print production, gradually brought down the costs of image reproduction in books and journals, making it possible to reproduce more images and higher-resolution color images. When in 1990 the *New England Journal of Medicine* introduced its "Images in Clinical Medicine" section as a regular feature of the journal, it was not only because images are, as the editors put it, "an important part of much of what we do and learn in medicine," but also because images could be more readily reproduced and discussed in lay, professional and scholarly journal essays. As digital photography and video offered interpretive and discursive tools to laboratory studies researchers in sociology and anthropology of science, theory and method changed to incorporate audiovisual modalities as more than tools of documentation.

An important component of my interpretation of the visual and digital turn is that much of the research about visual practice in laboratory studies using recorded video data has subscribed tacitly to an anachronistic realism in which the camera records data meant to reflect the profilmic real of laboratory conditions. Science studies scholars who use cameras and video tend to use the camera to document with straightforward and nonreflexive techniques. The Marxist and feminist materialisms of the 1970s and 1980s conveyed to the visual turn in science studies a focus on body-technology relationships and the material conditions of embodied practice reflexively taking into account the conditions of research and the emplacement of the science studies researcher. In visual studies, the role of the researcher herself in the nexus of production and observation has always been a critical point of concern. Not so in science studies, where visual or audiovisual data collection tends to be unreflexive. The reflexive traditions in visual anthropology and in feminist performance in art practice have had little impact on science studies. The current turn toward materialism and ontology in science and technology studies can be interpreted in continuity with the sociology of scientific knowledge in science studies, a trend that was itself in tandem with apparatus studies within film and media studies during the 1980s and 1990s. But attention to the apparatus in both eras has not extended to the apparatus of the science studies researcher. My agenda here is to introduce an example of just such a reflexive approach to documentation in visual science studies.

Finally, I wish to highlight a tendency toward skepticism of the visual in science studies scholarship then and now. This skepticism toward, or disavowal of, the visual is exemplified in Latour's "blind alley" comment. This disavowal is captured as well in the phrase "images are not the only truth" uttered by a scientist interviewed in a sociological study of imaging, as I shall discuss ahead. The comment stands in stark contradiction to the conditions of the workplace in which scientists and clinicians were seeing a dramatic ascent of digital imaging and graphics technologies, a turn in the practice and markets that occurred during the same decade of science studies' visual turn.

Like any contradiction, this one invites interpretation. We may discern, through interpretation of visuality in the context of materialism, exactly what is produced in the course of this disavowal. If visual studies is a misleading undertaking for a science studies that seeks to know more about knowledge practices, then it would be useful to understand what visual studies might bring to a research agenda that does not organize itself primarily around interpretations of knowledge projects, but which instead observes and interprets the material and ontological conditions of experience, the conditions of inscrutability, and circumstances in which research is sustained in states of not knowing.

I begin, then, by asking not just how science studies scholarship has addressed images and the matter of the visual and/as knowledge, but when and how science studies has taken into its canon the feminist materialist methodologies and theories that gave rise to a visual studies that could account for subjectivity, complexity and the failure of knowledge projects.[7] One approach to this question would be to identify instances of explicit attention to the visual and its methods in science studies. One might, for example, track key collaborations and dialogues between science studies scholars and art historians or film and media scholars. However, I wish to draw attention not only to work that has adopted the objects, methods and theories of visual studies but also to that which has critiqued and dismissed the visual as a deceptive or misleading focus for the science studies field. There has been some permeability between visual studies and science studies in their decades of formation, and skepticism and even repudiation of the visual have been points of consideration in both fields. The question of the limits of the visual field that has been an important aspect of science studies work on the visual is a central issue in visual studies as well. But science studies has not embraced the strong tools of critical analysis offered in visual studies. Likewise, visual studies has not been very open to the questions of a field that largely has been organized around objectivity and knowledge. Petchesky's essay was part of a vital conversation and group of publications and journals devoted to feminist theories, methods and issues in science and medicine. Reproduction was a dominant concern. The context included the Nordic collective Finnrage and the groundbreaking work of anthropologists and sociologists Rayna Rapp, Emily Martin, Faye Ginsburg and Sarah Franklin, and Adele Clarke. Despite Petchesky's early intervention and the body of work in dialogue with her essay during its time of publication, science studies has proceeded, for the most part, as if pleasure were of no consequence to scientific practice and as if subjectivity were of marginal concern.

From its beginnings, feminist visual science studies has taken as its object the materiality of embodied experience and the more-than-representational. If images and visuality, in science studies, are a kind of perverse object in the field (Latour's "peculiar")—something that has drawn our attention away from the important matter, whether that matter be the real, the multisensorial complex in which vision is embedded or the relational conditions of actors in the network—then it is precisely to the visual field that we should look. Recognition and denial are always mutually constitutive. There is much more to glean from images than the matter of objectivity and visual knowledge. Feminist visual studies into biomedicine, I will propose, has understood visuality to be embedded in a world of experience and practice conceived through theories that preclude reduction of visuality to images, the sense of sight and the merely representational.

But to return to feminist visual science studies and the matter of the body: Writing in 1994, Dorindra Outram, historian of the body and the French revolution, expressed concern that the matter of the body was being reduced to the issue of representation even as she noted the escalation of imaging practices in the current milieu. She began, "We live in a society which is saturated with images, which devotes more economic resources to the production of images than has any previous society, and which also carries out more of the political work of control and restraint through the conditioning carried by images, than ever before" (Outram 1994, 130). But in her investigation of the political culture of the body, she also asked whether the history of the body was not in fact being mistaken for the history of its representation. Her question was echoed in a different context more than two decades later by Karen Barad, who asked, "What compels the belief that we have a direct access to cultural representations and their content that we lack toward the things represented?" (Barad 2007, 132).

Outram and Barad missed the point that the turn to visuality and representation in feminist theory was precisely through a materialist theory that understood representation to be always already in relationship to an apparatus in which the body and its representation were coproduced. It may be true that the question of visual culture in science had, in past decades, been formulated by some science studies scholars primarily around the notion of the primacy of the visual register, representation and visual data. But much of this work did in fact fully embrace the matter of technology, materiality and multisensory engagement. *Matter* was at play in visual theory all along.

No sooner was visuality identified as a legitimate object of study in a range of fields touched by the visual turn than the scientific image was entered into a field of debate about visuality's place in the study of knowledge systems. Cautionary positions in relationship to the late 1980s "visual turn" in science studies have included taking issue with the segmenting out of visual things (representations, images) and practices (observation, looking) apart from other sorts of things (organs and molecules, instruments and spaces) and practices (talking, reading and writing, gesturing and touching) and the collapse of the distinction between representations and the things for which they stand. Even before the visual turn of the late 1980s, philosopher of science Ian Hacking observed that observation is overrated, pointing out that in scientific experimentation, "the test of ingenuity or even greatness is less to observe and report than to get some bit of equipment to exhibit phenomena in a reliable way" (Hacking 1983, 167). Hacking shifted his readers' attention from visuality as observation (seeing, sensing, perceiving) to the technological means of production and reproduction through which observation is enacted. This shift was in keeping with the Althusserian materialist turn to the apparatus associated with a range of disciplines and schools of thought during the early 1980s, a turn that was evident in the sociology of scientific knowledge (SSK) within science studies and in apparatus theory in film studies.

In light of Hacking's point we might consider the late twentieth-century rush, in cognitive science and neuroscience, for example, to magnetic resonance brain imaging as a source of knowledge about human behavior. To what extent are laboratory brain-imaging practices oriented toward generating new knowledge about brain and cognition, and to what extent is this work oriented to the task of getting a million-dollar bit of equipment reliably and consistently to show us something researchers across a range of laboratories might agree that they see? But to put Hacking's old insight about observation to work, we would need to understand the extent to which the technological enactment of machine vision is in fact an iteration of observation in the context of a spectrum of practices, rather than conceptually to diminish the role that observation plays in this context. Observation is not so much overrated as it is misapprehended as a unique modality, when in fact it is situated in a multimodal complex of technologies and practices.

Lynch proposed in 1994 that not only observation but also representation is overrated. He took to task this term in its status as a key epistemological concept among antipositivist philosophers of science (Lynch 1994). His target was the influential work of actor-network theory (ANT) generated by Bruno Latour, Michel Callon and others (Callon and Latour 1981; Callon, Law and Rip 1986), which he faulted for merging the political sense of the term representation, an authority delegated to speak and act on behalf of a constituency, with the semiotic concept of a sign, also understood as speaking on behalf of its object. Lynch argued that the merger, intended to allow an unraveling of lines of force and networks of power, had in fact fostered a thematic obsession with representation, obscuring the everyday activities of science that should in fact be subject to a more detailed, "exploded" examination of how representations are in fact used in a variety of sometimes contradictory, complex and mundane ways with other modalities of practice, and not always with meaning or knowledge as significant end points. A point that Lynch missed, however, is that the "obsession" with representation was not isolated to science studies scholars of the era, but plagued scientists and the public as well. Feminist visual studies scholars of the era were engaged in analyzing precisely this matter of an obsession with the visual, identifying it not as a mistake but as a mode of interaction in everyday practice across numerous professional and everyday cultures. To reduce the matter of visual obsession to an artifact or error of the analytic process is to dismiss one of the chief characteristics of sensory engagement (including and beyond the visual) in any domain of social practice, which is the tendency of the sensory to immerse or to carry those engaged in it into sentient states of feeling, of which pleasure is one possibility. What feminist visual studies saw that Hacking and Lynch missed is that if the visual was 'overrated' it was because it was seductive and exciting to scientists and STS researchers alike. To dismiss the visual as overrated is to overlook the role of pleasure as an important factor in scientific process

In the 2000s, science studies ethnographer Morana Alač adapted the ethnomethodological approach associated with Lynch (1985) and the engagement with technology in practice of which Law (1986) is a well-known example to the study of the experiential context through which laboratory work with images is performed. Alač takes Latour's (1986) notion of "thinking with the eyes and hands" into the realm of a study of brain scan interpretation and usage in cognitive science, showing that the activity of producing new knowledge entails multimodal semiotic work involving not just image and gaze but also gesture and talk. This approach is captured in her book title's reference to "handling" digital brains—a naming of process that foregrounds the materiality of semiotic and representational engagement. *Handling* moves us past the notion of the digital as a virtual world restricted to images accessed only through the sense of sight, drawing our attention to the multisensory process enacted in distributed networks in which researchers are physically copresent and interact with and through images in the laboratory, clinic and classroom (Alač 2011). As something to be handled intersubjectively and multimodally, the digital image is just a piece of the action and the production of knowledge just a part of what transpires in laboratory interactions. Rather than reducing the visual "blind alley" to a turn the science studies scholar might better avoid, Alač identifies the visual as a component of the complex of practice in which touch and proximity are also at play in the negotiation of digital matter. However, her attention to the material relations of the laboratory does not extend to the more broadly cultural—for example, the experiential aspects of embodied action and materiality that involve identity beyond that of the professional subject, or feelings outside knowledge, are not key foci in her investigation.

The view that knowledge should be decoupled from the image was succinctly captured by one of the brain-imaging scientists interviewed by science studies sociologist Anne Beaulieu: "Images are not the only truth," her scientist-research subject cautioned (Beaulieu 2002), suggesting to Beaulieu that an iconoclastic outlook had emerged among the scientists she studied who worked with MRI systems. The perspective is that we should not lose sight of information and the matter of numbers, the data crunched to make images. For this researcher, the image is in fact a red herring in a system that is a product of the quantitative tradition that would have us put our trust not in pictures but in figures, information that brings us back to the network and the relational and process-driven nature of knowledge systems (Beaulieu 2002, 2004; Porter 1996).

Beaulieu's researcher may be correct; however, that image is a placeholder for information does not diminish the fact that images hold a particularly compelling place outside the context of information. The image may very well be decoupled from its place in a scheme of accurate data representation and authority, as Petchesky's early analysis of fetal images and their place in gendered cultural pleasure showed. The fact that MRI has "imaging" in the very branding of the technique should tell us something about the cultural place assigned to images and the visual, even if images stand for or point to some more significant register of meaning. I propose that rather than moving away from images and representations to follow the things they point to (data, objects in the field), or to study the technologies and flows of knowledge (through information, numbers, networks), we might focus on the nature of the disavowal of the small "i" (the image) in a field where work involves using such an expensive bit of technology, and a technique that takes the "I" as a definitive term (MRI) in a global industry (medical imaging) that doubled from $10 to $20 billion in the first decade of the 2000s.[8] The perceived iconoclasm in the disavowal of the image is even more remarkable when we consider the complex cultural work involved not only in translating numbers into indexical images but also in transforming the indexicality of the image into a source of iconic meaning so distracting as to incite professional condemnation of the form. Iconoclasm conjures vividly what it disavows. If the image is nothing more than an index and a poor one at that, then why would it persist as an expensive and widely used element in an array of medical diagnostic and treatment systems that are "really" about indexicality through data, measurement, numbers, and information to be gleaned for knowledge? In fact these systems, whether in the form of numerical data, graphs, or images, offer evidence of materiality and place.

The distinction between indexicality and iconicity is at stake in this interpretation of the MRI as never merely an image. The focus on images as indexical objects in twentieth-century scientific research has been fairly substantial since Lynch and Woolgar's 1990 *Representation in Scientific Practice*, a collection that demonstrated the use of semiotics and laboratory study to consider work with images in laboratory practice and in the broader discursive networks in which scientists practice (labs, the field, or the context of professional journal publishing, for example). Invoking the cinema studies field emphasis on the temporal and durational nature of the image as a unit in the situated context of practice, Latour has stated, "There is nothing to see when we do a freeze-frame of scientific and religious practices and focus on the visual instead of the movement" (Latour 1998). However, at the time science studies theory did in fact have at hand a significant repertoire of work on the moving image to consult about the static frame in its uses for interpreting the nature of movement in forms such as networks. We need only look to the precedent of Ray Birdwhistell to understand that the moving image at midcentury was already a scientific modality developing in tandem with the twentieth-century map and other instruments and forms of mobility (see, for example, Davis 2001). Breaking with the medium specificity of visual studies, Latour also suggested, "If one now translates semiotics by path-building, or order-making, or creation of directions, one does not have to specify if it is language or objects one is analyzing" (Latour 1997). Representations and things are provisionally leveled when released from the perception that they internally harbor material conditions that delimit their potential activity. In this Latour echoes the perspective of Umberto Eco, for whom the iconic image (read the photographic image) is, like the verbal sign, "completely arbitrary, conventional and unmotivated" (Eco 1970). In light of much of the canonical work in the history of photography addressing the perceived indexicality of photographic records (Bazin, Barthes, Doane), Eco's understanding of the photograph as iconic reads as an unusual aberration for someone steeped in the very linguistic systems that moved photographic theory away from its emphasis on (mere) pictorialism. Representations have never been understood as merely static pictures, insofar as images embody forms of mobility and relationity in every specific case of their use. Moreover, a map is no more or less equivalent to a static picture than, say, a chart, a graph, a segment of film, or a piece of code.

Latour developed this notion of the static frame in a manner that avoided the language of the index, emphasizing the deictic capacity of pointing and context in his instructions on "How to Be Iconophilic in Art, Science and Religion," an essay in which he uses art history to suggest an "iconophilic" way to interpret the use of maps. Whereas the iconoclast dreams of unmediated access to truth (an absence of images), the iconophile focuses on the movement of mediation and the provisional nature of the image in a network (Latour 1998, 421). To make his point, Latour draws an analogy between two images. The second is a painting by Fra Angelico, in which one hand of an angel points down toward an empty tomb while the other points up toward a resurrected Christ. The angel's gaze is leveled at a group of apparently bewildered women. The first image is a photograph in which soil scientists "dominate" a landscape through deictic gestures, manipulating, pointing and gazing at a map. The scientists' activity is oriented toward an elsewhere that we might understand as outsdie the frame of the image (recognizing this map is in fact a representation of a distant landscape that rests out of the scientists' own visual range as well). Latour suggests, with respect to indexicality, that the map is an entity that takes on material significance when put into motion precisely through the movements of gesturing angels (the scientists) and the circulation of immutable mobiles (which include the table, the hands, the eyes, and the map itself). For Latour, a world where we have angels and immutable mobiles circulating and that is accessed through prayers, discipline and aligned intermediaries is better than the "horrendous culture in which the poor angels are harnessed to do the work of instruments, accessing a world beyond and carrying blank messages on their return" (Latour 1998, 438).

The disclaimers about the status of images, representations and observation I have discussed thus far could be taken, together, to suggest that the visual turn was a short-sighted and narrow detour on the way to sensory multimodal research about the network; that perhaps we should focus, after actor-network theory and its critics, on networks of agency and practice or, after ethnomethodology, on the complex of materials and sensory and technical modalities that structure the field of embodied situated practice in the laboratory and the field. Latour's enthusiasm for an embrace of iconophilic movement with images demonstrates an interesting problem about the regard of medium specificity, an issue that has been a major concern for the visual studies field in an era of convergence: No sooner are representations and representational practices considered than they are entered into an analytic framework in which they provisionally cleave to other matter—other bodies; the objects, spaces and technologies to which those bodies point with fingers, gazes and utterances; other images and imaging technologies; other representational practices. Like RNA viruses, images function in a cloud of quasi-species in which the transformation of matter and meaning is constant and diffuse, eluding critical reception. The image loses its physical distinction in a sea of mobility, referentiality and mediation.

I want to suggest that in fact the critique of the obsession with representation and the iconoclasm identified by Beaulieu's research subject (2002) in fact point to a larger problematic concerning imaging: that of the disavowal of attention to images and visual data precisely because they "shoot blanks" and lead one down "blind alleys" like that quintessential space of the early ultrasound image, the womb. The "blank message" deserves close consideration along with the dead-end optical space of the visual field. As Jimena Canales (2010) shows in her book *A Tenth of a Second*, the moving image's static frame is an index of becoming and not an iconic placeholder for temporality. It does not point to the vacuity of stasis in a world that operates through networks. The static frame is a record of an increment of marked time toward the always hoped-for appearance of the real that representations guarantee but don't deliver in any fullness of being. Imaging systems may indeed shoot "blanks," but blanks are significant in their failure to deliver the imagined referent at any given or desired point in time, as Petchesky's article on fetal imaging so cogently showed about the received timing of fetal life and citizenry relative to the frame-image of its nascence. These entities, the visual and the static frame, require attention specifically from visual studies in order to grasp the nature of the materiality of images in relationship to the materiality of imaging as a process and the temporal conditions of being. I want to suggest that we look with specificity at the problem of "blank messages" and the stakes of invisibility and sustained "not knowing," and of not having a clear indexical picture-referent relationship, in research and clinical practice.

A familiar story in visual studies and criticism of the late twentieth century is the narrative of the bankruptcy of visual enlightenment. In these accounts, the visual quest for knowledge is shown to have yielded blank messages. Opening up the body to observation and imaging has resulted not just in new knowledge, cure or the prolonging of life but also in the production of blind spots, erasures, diversions and dead ends. The postgenomic era is marked by counternarratives of the failure of genomic science to yield the knowledge that was anticipated in the quest for genetic code. The *New York Times* quotes David Goldstein, a population geneticist, in a 2008 article that remarks upon the failure of the Human Genome Project to identify variant genes that predispose people to diseases such as cancer—one of the primary motivations behind the $3 million project that bridged the centuries: "It's an astounding thing that we have cracked open the human genome and can look at the entire complement of genetic variants," Goldstein remarks, echoing a point made many times over by scientists celebrating the genetic-informatic turn. But he completes his thought with a compelling irony: "and what do we find? Almost nothing. That is absolutely beyond belief" (Goldstein, quoted in Wade 2008). One could take from this point a sense of the failure of the enlightenment knowledge project—or a sense of the importance of grasping "failure" as a condition of practice in which blind alleys may be the sites of scientific practice more often than researchers realize. They require a method for better negotiating this sort of a structure, and not simply a way out to the open network with all of its affordances and modalities at hand.

The point, tacitly expressed by Goldstein, that we now know that "to see is not to know" coincided with a turn away from epistemology to ontology as a primary object in twenty-first-century science studies. Catherine Waldby, in her book-length study of tissue culture, published in 2006, proposed that those who work on the concept of body image have been too preoccupied with the register of the visual at the expense of introceptive data. In the twenty-first century Waldby's research focus shifted from medical imaging (her previous book was about the Visible Human Project) to biological matter and the transfer of tissue fragments. Waldby's shift was part of a new materialism in science studies, a move from representations to bodily matter. The renewed focus on the body as matter and on the material conditions of mediation rather than on representation, image and knowledge recalls the materialism of an earlier era—that of the sociology of scientific knowledge (SSK) focus on the apparatus, for example, and that of the very Marxist feminist materialsm that was critiqued for its troubling essentialism in the 1980s. In what is also a turn away from the ANT tendency to level matter in the interest of following the network, the new materialism posits that the distinct qualities of matter must be foregrounded and addressed in their mutable specificity (Bennett 2009; Coole and Frost 2010). Medium specificity, design and the instrumentation of matter—its physical negotiation—have come to the foreground again in work that might be characterized as targeting matter as mutable and immobile, in the sense that the strategy is to stay with matter and do the slow work of its critical articulation, its intra-specificity and its internal complexity rather than its intersectionality and disparities with contiguous forms. The point is not that objects don't move around the network, but that the materiality of objects may change internally and in slow time, requiring attention to the specificity of their conditions of temporality and materiality at the scale of the minute and the invisible.

Barad's critique of representation is that language has been granted too much power, that every "thing"—even materiality itself—is turned into a matter of language or some other form of cultural representation. But the turn to representation, which cannot be reduced to the linguistic turn and which was also very much a materialist turn involving attention to the apparatus, was meant precisely to address the limitations of the linguistic turn and human consciousness in their ability to further the agendas of addressing materiality and the more-than-human features of the material world. Materialist work on the apparatus, in Marxist and feminist film studies, refocused attention to the materiality of film experience, without ignoring the status of the film "text" as always also entered into exchanges of meaning and intra-action within the apparatus. In its anti-humanist critique of representational realism, feminist film theory offered a powerful precedent to the focus on the more-than-human that has been a rallying point for the new ontology in science studies.[9]

I am in agreement with Waldby in her turn to the materiality of the body as an urgent matter where health and illness are concerned; however, I want to emphasize that tissue culture has emerged precisely as a *medium*, as she amply shows. Tissue culture is strongly marked by an ethos of temporality and visibility, an ethos that I and others have suggested is cinematic in its material nature. As Hannah Landecker (2005) shows, nineteenth-century cell microscopy and twentieth-century cinemicrographic imaging did not simply *represent* cell research. These technologies served as direct means of experimenting with the physical and temporal dimensions of cell growth that have transformed human tissue into a viable industrial medium for laboratory work. As both Waldby and Landecker show, human cell lines have emerged as patentable media that trade on their material and indexical relationship to an original (Waldby 2002, 2006; Landecker 2007). Tissue culture is a corporealized transference medium of intellectual property in the global biopolitical arena. This is the case in part through recognition of physical consistency between the "original" cell line and its replicants, which require medium consistency and specificity—cross-contamination with other cell lines ruins the medium. The regard of tissue as medium is a concept that rests heavily on the paradigms of visuality and mediation: Cells emerge in an indexical relationship to life. This quality of indexicality consists not in mobility (the cells are not strictly immutable mobiles that convey meaning) but in a more radical specificity of material qualities that convey from substance to substance, in a mutability that requires contact and a small scale on which movement occurs. For indexical meaning to convey, a particular substance has to convey with a closeness and a particularity. This is different from the work of the immutable mobile, which is neither medium-specific nor small in its scale of movement. Tissue as medium must be understood as more than a provisional frame that points to remote phenomena, to something else; it must be understood in its internal relations and in its intimate proximity to other forms on multiple levels to perform as it does.

I wish to extend this interpretation of the materiality of tissue as medium back to the passé notions of image and representation, in order to gesture back to the discourse on representation launched by Mulvey, which attributed materiality and material specificity not only to both the film medium and its referent but also to the complex field of actors, agents, objects and technologies engaged in the field of action. Rather than turn away from images because they shoot blanks where the matter of the referent is concerned, I wish to suggest that we look closely at images in their indirect, complex, contextual, proximal relationship to other sorts of matter. Barad's rejection of representation cited earlier is based in an older model of thought that places illusionism (or representation) in opposition to materialism understood to be more authentically a quality of the referent (or, the profilmic body or object). But as Maria Walsh notes in a review of a recent book by Laura Mulvey, "In the current era of technology, these oppositions [between representation and referent, for example] are becoming less distinct" (Walsh 2006, 1). For Walsh, these pairings give rise to dialectical uncertainties rather than dichotomies such as profilmic and representation, matter and its representation. As the case of tissue culture shows, the medium is not always or solely referential. I further suggest that pairings such as "image and referent" are rich ground for investigation of the material conditions of representation as a component of the matter at hand in any situated practice.

Sensory ethnography, with its technologically facilitated attention to the condition of embodied looking practices in the field and also in the activity of the science studies researcher, is a potentially productive method for the project of attending to the place of the visual in a materialist approach that addresses the multisensory and multimodality in the variegated digital field of science studies research. I believe the dichotomy between representation and the real is a straw man of current theory. Most who work in visual theory and in theory-practice work on science and technology do not believe that the binary model of representation is operable in most instances of scientific practice; it certainly does not capture the nature or work of images in contemporary scientific imaging pratices. Pursuing observation and interpretation of the internal material conditions of the many "blanks" and "blind alleys" that make up the history of scientific practice is an endeavor that may be pursued through a range of methods. In the remaining pages below I focus on materialist theory-practice work in sensory ethnography in which the visual register becomes a means through which to engage the complex multisensorium and the multimodal ways in which becoming and experience unfold in the space of the clinic.

MATERIALIST THEORY IN VISUAL FORM

Theory is performed not only in writing but also reflexively in audio, visual and audiovisual form. The Austrian sociologist and videographer Christina Lammer has been engaged for three decades in research that performs videographic observation and theorization of intersubjective experience in the biomedical context. She offers what the British sensory ethnographer Sarah Pink calls a multisensorial approach to understanding one's subjects in the course of observing their situated material practices (Pink 2009). It is to Lammer's work that I turn to provide an example of a practice that is radically materialist and which engages with the visual to discern and articulate the intrastitial conditions of the clinical apparatus. A sociologist with a long-standing engagement with the video recording of interaction in clinical and surgical settings, Lammer studies interaction at the microlevel, using some of the techniques familiar to science studies ethnomethodologists. Lammer's video observations foreground the intersectional experience of medical professionals and patients, which she documents at close range. Her camerawork takes into account factors such as whether a patient is conscious during a procedure, how physicians interact in teams with one another and with nurses and other staff active in the procedure, and how technologies and equipment, from gloves to implements, come into play. Lammer has been granted a rare degree of access to surgical settings at the Medical University of Vienna, where for over a decade she has engaged in a series of ethnographic studies of surgery. A sociologist by training, Lammer works in a tradition of ethnographic interpretation of clinical interaction with a few notable precedents. Her approach, like that of the visual anthropologist Richard Chalfen (Chalfen 1998, Chalfen 2000), emphasizes the phenomenological and embodied experience of human subjects immersed in illness and treatment experiences. Well known for his strategy, developed in the 1970s with the visual anthropologists John Adair and Sol Worth, of handing over the camera to research subjects (Adair and Worth 1977), Chalfen has been engaged in the use of the camera as a participatory tool in the negotiation of the health care apparatus for many decades, most recently as associate scientific staff at Boston Children's Hospital. Lammer's work with the camera in clinical settings through the Medical University of Vienna and other sites is distinct from Chalfen's in a few important ways: Whereas Chalfen offers the camera to his human subjects, Lammer maintains control of the camera, though of late she has offered her subjects pencils and brushes and documented their processes of drawing. Drawing and painting are means through which her subjects express and interpret their intrapsychic and affective experiences in the clinical process, whether as caregiver or patient. Texture, detail and proximity and the internal components of the clinical experience are strong concerns for Lammer. Intra-action and intra-acting agencies, concepts introduced by Karen Barad (Barad 2007, 206), are useful terms through which to interpret the scale and degree of proximity with which Lammer engages her subjects. Barad writes of the ontological inseperability of intra-acting agencies for which the apparatus is the material condition of possibility for human experience. A fair number of visual ethnographers of science and medicine and documentary filmmakers have studied interaction in the hospital setting, but few have gained the level of access Lammer has been provided to the entirety of the surgical process itself, from the clinical consultation through surgery and into postoperative recovery and reconstruction. Most importantly, Lammer focuses on the subjective experiences of her subjects at what we might call the intrastitial level, borrowing from Barad's notion of intra-action here to bend back in upon itself the more familiar concept of the interstitial (the empty space between spaces full of matter) in order capture this sense of documenting the internal material activity and viscerality of the empty in-between that makes life possible. Lammer proceeds not simply by recording what her subjects say and do intersubjectively, but by showing us the details of their embodied experience through the apparatus of the clinic, including multimodal interactions under anesthesia in close ups that bring us to these points of intrastitial materiality. She reveals nuances of emotion and intimacy in intra-action, putting aside questions of knowledge, language, and larger-scaled interaction (between body and technology, for instance) in favor of discerning affect and phenomenological experience as it unfolds at a much more internal scale. We see hands inside bodies, multiple hands of multiple subjects acting in unison as the apparatus of the surgery refuses distinction among different human agents who together constitute the apparatus and whose parts are inseparable from it. Lammer's time frame is broad (she follows treatment from diagnosis through postsurgical visits), but her scale of focus takes us into the microlevel of interaction and expression where individual subjectivity is no longer relevant or at stake in the distributed micro-level enactment of the material process.

Hand Movie 1

Included in the Montreal OBORO exhibition Auto/Pathographies (2012), the video *Making Faces* is a work of 2012 in which a young girl performs facial exercises in preparation for facial reconstructive surgery. Lammer's camera shows in detail the girl's subtle movements and expressions. The close-up footage sustained over a long take and projected on medical gauze invites the spectator to sustain a stare, not to be intrusive or voyeuristic but to better understand facial reconstruction as a process that involves much more than the surgery itself and entails the training of muscle memory. The spectator is brought into a sustained engagement with the process through which the child must confront herself in the mirror and teach her new muscles, adapting them to the changes of the look and feel of her own face. Few visual ethnographers have taken the approach of observing bodily movement and intra-action at the extremely close range and scale of interaction that Lammer provides without reducing this kind of performance to the expression of meaning and named emotions (as in the case of the work of Paul Ekman, for example, where meaning is the bottom-line concern in his videowork documenting facial expression). Lammer's use of a long shot capturing a seemingly benign process is common in the observational documentary tradition of direct cinema, a style and movement variously associated with directors such as the Maysles brothers, D. A. Pennebaker, and David and Judith MacDougall (MacDougall 1998; Grimshaw 2009), however, in the case of that tradition sync sound is essential to the unfolding of human experience. Here we have the sustained look of the *Screen Tests*, Andy Warhol's silent close-up film portaits made of 7-minute continuous shots. I have made the suggestion that the sustained close-up of a young girl does not invite voyeuristic or sexualized looking. Rather, the camera invites a gaze that is both clinical and empathetic. The footage invites a clinical gaze in its direction of the spectator's attention to the minutiae of the facial exercise, allowing the spectator to stare, to collect information in the sustained and cool manner conveyed in the Barthesian notion of "studium" (Barthes 1981). This is a kind of civility of looking that creates interest and invites recognition not of the subject per se but of the operator—the one who has filmed this face, and the one who actively looked through the viewfinder for the duration of the take. Rosemarie Garland-Thomson captures this comportment of cool looking well in her book *Staring: How We Look* (2009). Describing the practice of staring at faces, Garland-Thomson highlights the role of the face as the unclothed portion of the body to which one looks to collect information (the project that "studium" describes so well). However, Garland-Thomson also notes the degree to which the stare can offer the spark of recognition, and the extent to which it may position the spectator to serve as witness not just to information but also to feeling. These aspects of feeling and witnessing are better captured in the Barthesian notion of the "punctum" (Barthes 1981; Garland-Thomson 2009). Through the instruments of the hand and the camera Lammer offers an empathetic relationship of looking closely and with care, even as the image offers data. She coins the term empathography to describe her process, bringing together empathy with videography.

Lammer's empathography also involves documenting the subtle, almost imperceptible ways that surgeons and others in the operating room work together not only with obvious empathetic feeling for their patient but also with a close material connection among themselves on the most minute scale of interaction such that the space between bodies becomes full with intra-action. Moving the camera in close to the surgical field and to the face and body of the human subjects under treatment, Lammer shows that the hand of the surgeon is not simply a precision tool for getting things right. It is also an empathetic and creative extension of the feelings—the hand of the surgeon that feels for the body of the patient, and that works with feeling along with other hands that operate upon the body of the patient. There is a tremendous tenderness in the relationships of hands belonging to different bodies documented by Lammer's camera as they work in concert on an unconscious subject's body.

Lammer's *Hand Movie 1*, a short work in a series of three hand "choreographies" of the surgical process, opens with an extreme close-up on a pair of hands folded contemplatively and covered in sterile gloves tucked into the arms of a surgical gown. The two thumbs support one another, bouncing together in rhythm as if performing in dialogue the inner thoughts of the surgeon as s/he plots the first steps (a word that suggests movement and which we take from the action of the feet and not the hands). It is as if the surgeon is limbering up for the performance ahead, stretching the fingers as a pianist might do before an important performance for which there has been no time to warm up. But there is never the opportunity for rehearsal. All cuts are a live performance.

These two hands are framed in chiaroscuro, a uniformly shaded background that holds some folds of the sheet covering the operating table—and I pause over the word table here. The intimacy of the shot makes me want to call this table a bed. What does it mean that we choose the word "table" to describe the place where our unconscious body is laid to rest for the duration of something as personal and intimate as a surgical operation? As Sara Ahmed (2006) notes, a table is a place where we are oriented in consciousness to write, to eat, to display and contemplate things—or, it is a place where we classify parts into taxonomies, as in a table of elements, or a table of illnesses. The body rests upon a sheet as if in the safety of a bed where we would fall unconscious in a different way that also involves trust and private intimacy about being touched in that space of unconsciousness.

Before *Hand Movie 1* breaks into a sea of action, a single hand reaches into the frame and places itself steadily and firmly, gently, on the chest of the unconscious body. It is a strange and moving moment. Such a gesture would never be made by a medical professional toward a conscious subject; even in clinical exam, the hand of the clinician would never be made to rest, even in reassurance, over the heart. But in this context, with the patient unconscious and about to undergo surgery, there is no mistaking that the gesture is an expression of care. The image is touching because it gives a strong sense of the caring hands that receive this body in its state of unprotected vulnerability. At this close range, the view of the reassuring hand works almost like a point of view shot. In this close-up of the touch, we are invited to feel, as if for the patient undergoing surgery, a sense of reassurance, a confidence in the many hands that will perform in concert. We might also sense what this hand feels—the cool, soft tension of the flesh where the cut will be made, the warmth that radiates from the slumbering body, and the calm pulse that will remain so despite the reparative violence that will ensue.

Gloves are a predominant element in this movie. They conceal, but they also reveal action. We can see the outline of the surgeon's thumbnail, but too the glove draws our attention to its paradoxical role as mediating tool: Thin and supple yet impermeable, the glove makes possible a kind of intimacy and direct contact by touch while shielding from any potential exchange of infectious matter. The surgeon can feel though his or her instruments the release of the tension of the skin as an incision is made. The gloves mediate the soft, wet warmth of organs while preventing the passage of traces of infectious matter from the outside world into the vulnerable tissue of the patient's body. The gloves also keep the fluids of the patient's body from seeping into the surgeon's pores. The gloves allow so much sensory intimacy, and yet they hold such a firm border against the mixing of matter. At this micro-analytic scale of seeing, we can also feel the difference between the warm, soft tissue of the living body and the cool, soft gel of the implant in its sealed plastic membrane, and the cool steel of instruments as they must feel against the vinyl skin of the gloves, and the tension of the thread as the needle pushes through the skin and is pulled, that faint sensation of friction as the thread is pulled through two thin layers of fabric in a reparative movement that brings closure.

What I am trying to emphasize in this descriptive account is how Lammer's sensory film draws out for us not simply what can be seen with this kind of close-up gaze, but also what is felt by the conscious human subjects in this field who perform with a high degree of multisensorial interconnectedness, allowing us to experience that field of sensation in its dense complexity. The materiality of the body and the materials of the surgical process come to the fore in this "blind alley" where knowledge and the matter of the image are always secondary to the sensorial data of the field, and to the affect that is both conducted and conveyed in these documents.

Hand Movie 2

In the second of her hand movies, Lammer shows us material from her observational footage taken during a surgery on the face of a six-year-old boy. We see in the shot two distinct surgical fields: In the background two surgeons work on the thigh from which flesh will be taken for the procedure on the face, while in the foreground two surgeons work on the face. Lammer has written that in this surgery, the face of the boy is small; so much activity, including her own work with the camera, must be done in a narrow space— eight hands working in the same field, or ten, if we count Lammer's own.

The observational phenomenologist David Sudnow, in his portrait of the craft of learning to play jazz piano, wrote, "I intend my descriptions as indications for how one might eventually speak methodically and rationally, if only crudely for now, when saying things like: the hand—in music, eating, weaving, carving, cooking, drawing, writing, surgery, dialing, typing, signing, whatever—this hand chooses where to go as much as 'I' do" (Sudnow 1978, 2).

In watching *Hand Movie 2*, the hand "chooses where to go"—that is, the hand of the surgeon moves as if with its own inner logic and rhythm that is not preceded by knowledge or intelligence, but that is intuitive and constitutive of meaning. But the hand always moves in concert with other hands—the surgical teams' hands move in ways that are deeply internalized and memorized as an intrastitial group activity, not by rote but with the flexibility of improvising together. The movie tells a close and personal story about intersubjectivity, where all of the communicating is performed not through speech but through the subtle, quiet microgestures of the hands—the small motor actions that perform in a field that requires a high level of awareness in eye-hand coordination not only about what one does individually but also about what one's partners are doing and probably intend to do, moment to moment. The success of an operation hinges on this intense microlevel of awareness and anticipation of what one's hands will do, and what one's partners' hands can be expected to do in concert with one's own. Accounts about surgery and subjectivity tend to emphasize the interaction between two subjects, as in narratives about the doctor-patient relationship, or the relationship of body to technology. I am sure we can all think of accounts we have read of the subjective experience of a medical encounter written or filmed from the standpoint of one subject—the personal standpoint of a doctor or patient, for example. These accounts almost always rely on words, either for the documentation itself or for the retelling. When the work of the hands of the surgeon is the focus, there is a tendency to fetishize them as the skilled tools that make the surgery a success. (It is a familiar convention to speak of the brilliant hands of a concert pianist or a painter, and the cultural fetishization of the hands of the surgeon captures this same sentiment.) Yet the surgery involves not just two but many hands. The closeness of the shots in *Hand Movie 1* and *2* invite us to see the intense sensory intimacy of this multiplicity of practiced hands performing as if prior to consciousness with the unconscious subject's body. No sooner do we see this ritual of the two hands at the start of the movie begin their work than the frame is filled with not just two but many hands, at times as many as eight, all working in the same field as one creative body. These hands reach into the field of the tight shot around the surgical site with collaborative gestures and movements that complete one another as tools pass between hands and as hands belonging to different bodies perform tasks in concert with one another. In rhythm and pace, the shots take on the dimensions of a concert, in that the hands work their instruments together with astonishing precision. Never do we see two hands getting in one another's way. Often it is impossible to tell which two hands belong to the same body, such is their practiced skill in working together as one many-handed body. One could say the team has come together as if all of one mind—and this without having to prethink any one gesture. But the more critical point is that this process seems as if to constitute mind, and not to derive from it. The routine of work with the hand of another has become internalized, so that the hand performs apart from the conscious thought-process of the individual surgeon or nurse. In this way, we might say that the patient is not the only one whose unconscious is actively on display in the operating room. Think, for example, about a shot in which the surgeon uses his or her dominant hand to pass off an instrument to the hand that waits to receive it from the other side of the patient's body. The hand that receives the instrument waits in readiness. That hand to which the tool has been passed remains poised, holding the instrument delicately but firmly and still in exactly the right position so that the surgeon can retrieve it again with no loss of time and no need to speak. This is not about consciousness, intentionality and meaningful communication: The patient is, of course, unconscious, but also the medical team works intuitively and mechanically in a long-rehearsed interaction that has moved beyond meaning and into the place of anticipation—the team, including the patient's body, is a body multiple, to borrow the term famously introduced by Annemarie Mol (2002) to describe the intersubjective experience of atherosclerosis in her biography of the disease.

CONCLUSION

I return to Barad's question: "What compels the belief that we have a direct access to cultural representations and their content that we lack toward the things represented?" Lammer's films suggest a third way that is neither a reduction of things to the meanings held in representations nor an ontological reverence for that which is more than representational. In Lammer's process, we can discern an engagement with the intra-play of matter through which the transformation of human subjectivity occurs prior to language and is constitutive of knowledge. The hand and the face are the two places we tend to look when we turn to the body to read its surface for expression and feeling. Emotions are understood to be "written" on the face, and it is the work of the hand that we associate with the crafting of meaning through gesture and the crafting of work through skilled manipulation, from the handiwork of painting to the handiwork of surgery. In this selection of Christina Lammer's hand films, the role of the image is as a tool for the contemplation of the multisensorial and affective dimensions of matter and the transformation of life in surgical experience.

NOTES

1. The term "science studies" is used throughout this chapter to describe the interdisciplinary social science, arts and humanities field that has formed around the study of science, technology, medicine and (more recently) information, and is represented by two scholarly societies: the Society for the Social Studies of Science (4S) and the Society for Literature, Science and the Arts (SLSA).
2. Petchesky cites the Mulvey (1975) essay in a footnote. The term "science, technology and medicine studies" will hereafter be shortened to "science studies." Field formation and naming are an issue worthy of discussion; however, there is not space to take it up in this chapter.
3. These two issues were edited by Paula A. Treichler and Lisa Cartwright at the invitation of founding editor Constance Penley, and were the basis for the anthology *The Visible Woman: Imaging Technologies, Gender and Science* (1998).
4. This is, of course, a generalization. As with all generalizations, one could point to exceptions. On the subject of standpoint theory in feminist epistemology, see Haraway (1991) and Harding (1991). For work on subjectivity in the sociology and anthropology of the body, see for example Akrich and Berg (2004), Lock and Farquhar (2007) and Biehl, Good and Kleinman (2007).
5. Jackie Stacey, James Elkins and I are among those who have actively participated in both fields since the 1980s. See Elkins (2007). See Jones and Galison (1998) for an example of a work that explicitly bridges art history and the history of science.
6. On this point see, for example, Beaulieu (2004) and Dumit (2003).
7. Although there is not space to develop this point here, Halberstam's notion of queer failure is a relevant potential source for feminist work in science studies (Halberstam 2011).
8. Medical imaging industry overview, the Audacity team, February 4, 2011, http://www.audacitygroup.com/2011/02/medical-imaging-industry-overview/.

9. For an account of anti-humanist, anti-illusionist feminist positions in film theory on the matter of representation and apparatus, see Kuhn 1990.

REFERENCES

Ahmed, Sara. 2006. *Queer Phenomenology: Orientations, Objects, Others.* Durham, NC: Duke University Press.

Akrich, Madeleine, and Marc Berg, eds. 2004. "Bodies on Trial." Special issue. *Body and Society* 10 (1).

Alač, Morana. 2011. *Handling Digital Brains: A Laboratory Study of Multimodal Semiotic Interaction in the Age of Computers.* Cambridge, MA: MIT Press.

Barad, Karen. 2007. *Meeting the Universe Halfway: Quantum Physics and the Entanglement of Matter and Meaning.* Durham, NC: Duke University Press.

Barthes, Roland. 1981. *Camera Lucida: Reflections on Photography.* Translated by Richard Howard. New York: Hill and Wang.

Beaulieu, Anne. 2002. "Images Are Not the (Only) Truth: Brain Mapping, Visual Knowledge, and Iconoclasm." *Science, Technology & Human Values* 27: 53–86.

Beaulieu, Anne. 2004. "From Brain Bank to Database: The Informational Turn in the Study of the Brain." *Studies in History and Philosophy of Biological and Biomedical Sciences* 35: 367–390.

Bennett, Jane. 2009. *Vibrant Matter: A Political Ecology of Things.* Durham, NC: Duke University Press.

Biehl, João Guilherme, Byron J. Good and Arthur Kleinman, eds. 2007. *Subjectivity: Ethnographic Investigations.* Berkeley: University of California Press.

Callon, Michel, and Bruno Latour. 1981. "Unscrewing the Big Leviathan: How Actors Macrostructure Reality and How Sociologists Help Them to Do So." In *Advances in Social Theory and Methodology: Toward an Integration of Micro- and Macro-Sociologies,* edited by Karin Knorr-Cetina and Aaron Victor Cicourel, 277–330. Boston: Routledge and Kegan Paul.

Callon, Michel, John Law and Arie Rip, eds. 1986. *Mapping the Dynamics of Science and Technology.* London: Macmillan.

Canales, Jimena. 2010. *A Tenth of a Second: A History.* University of Chicago Press.

Rich, Michael, Steven Lamola and Richard Chalfen. 1988. Video Intervention/Prevention Assessment (VIA): An Innovative Methodology for Understanding the Adolescence Illness Experience (with M. Rich and S. Lamola) in Journal of Adolescent Health 22:128

Rich, Michael, Steven Lamola, Jason Gordon, and Richard Chalfen. 2000. Illness as a Social Construct: Understanding What Asthma Means to the Patient to Better Treat the Disease. *Journal of Quality Improvement* 26(5): 244–53.

Clarke, Adele E. and Moore, Lisa Jean. 1995. Clitoral conventions and transgressions: graphic representations in anatomy texts, c1900–1991. *Feminist Studies* 21: 255–301.

Coole, Diana, and Samantha Frost. 2010. *New Materialisms: Ontology, Agency, and Politics.* Durham, NC: Duke University Press.

Coopmans, Catelijne, Janet Vertesi, Michael Lynch and Steve Woolgar, eds. 2014. *Representation in Scientific Practice Revisited.* Cambridge, MA: MIT Press.

Crary, Jonathan. 1990. *Techniques of the Observer: On Vision and Modernity in the Nineteenth Century.* Cambridge, MA: MIT Press.

Daston, Lorraine, and Peter Galison. 1992. "The Image of Objectivity." *Representations* 40: 81–128.

Davis, Martha. 2001. "Film Projectors as Microscopes: Ray L. Birdwhistell and the Microanalysis of Interaction (1955–1977)." *Visual Anthropology Review* 17 (2): 39–49.

Dumit, Joseph. 2003. *Picturing Personhood: Brain Scans and Biomedical Identity.* Princeton, NJ: Princeton University Press.

Eco, Umberto. 1970. "Articulations of the Cinematic Code." *Cinemantics* 1.

Elkins, James. 2007. *Visual Practices across the University.* Munich: Wilhelm Fink Verlag.

Garland-Thomson, Rosemarie. 2009. *Staring: How We Look.* Oxford: Oxford University Press.

Grimshaw, Anna. 2009. *Observational Cinema: Anthropology, Film, and the Exploration of Social Life.* Bloomington: Indiana University Press.

Hacking, Ian. 1983. *Representing and Intervening: Introductory Topics in the Philosophy of Natural Science.* Cambridge: Cambridge University Press.

Halberstam, Judith. 2011. *The Queer Art of Failure.* Durham, NC: Duke University Press.

Haraway, Donna. 2004. "Situated Knowledges." In *The Feminist Standpoint Theory Reader*, edited by Sandra Harding. New York: Routledge.

Harding, Sandra. 1991. *Whose Science / Whose Knowledge?* Milton Keynes: Open University Press.

Henderson, Linda Dalrymple. 1983. *The Fourth Dimension and Non-Euclidean Geometry in Modern Art.* Princeton, NJ: Princeton University Press.

Herzig, Rebecca. 2005. Suffering for Science: Reason and Sacrifice in Modern America.

New Brunsqick, NJ: Rutgers University Press.

Jones, Caroline A., and Peter Galison. 1998. *Picturing Science, Producing Art.* New York: Routledge.

Jordanova, Ludmilla. 1989. *Sexual Visions: Images of Gender in Science and Medicine between the Eighteenth and Twentieth Centuries.* New York: Harvester Wheatsheaf.

Landecker, Hannah. 2005. "Cellular Features: Microcinematography and Film Theory," *Critical Inquiry* 31 (4): 903–937.

Landecker, Hannah. 2007. *Culturing Life: How Cells Became Technologies.* Cambridge, MA: Harvard University Press.

Latour, Bruno. 1986. "Visualization and Cognition: Thinking with Eyes and Hands." In *Knowledge and Society: Studies in the Sociology of Culture Past and Present*, vol. 6, edited by Henrika Kuklick and Elizabeth Long, 1–40. Greenwich: JAI Press.

Kuhn, Annette. 1990. "Textual Politics." In *Issues in Feminist Film Criticism*, edited by Patricia Erens. Bloomington, IN: Indiana University Press.

Latour, Bruno. 1997. "On Actor-Network Theory: A Few Clarifications Plus More Than a Few Complications." Abstract for a paper. Accessed July 15, 2013. http://www.f.waseda.jp/sidoli/Latour_ANT_Clarifications.pdf. Last accessed 4 November 2013.

Latour, Bruno. 1998. "How to Be Iconophilic in Art, Science, and Religion?" In *Picturing Science Producing Art*, edited by Caroline A. Jones and Peter Galison, 418–440. New York: Routledge.

Law, John. 1986. "Laboratories and Texts." In *Mapping the Dynamics of Science and Technology*, edited by Michel Callon, John Law and Arie Rip, 35–50. London: Macmillan.

Lock, Margaret and Judith Farquhar, editors. 2007. *Beyond the Body Proper: Reading the Anthropology of Material Life.* Durham, NC: Duke University Press.

Lynch, Michael. 1985. *Art and Artifact in Laboratory Science: A Study of Shop Work and Shop Talk in a Research Laboratory.* London: Routledge and Kegan Paul.

Lynch, Michael. 1994. "Representation Is Overrated: Some Critical Remarks about the Use of the Concept of Representation in Science Studies." *Configurations* 2 (1): 137–149.

Lynch, Michael, and Steve Woolgar, eds. 1990. *Representation in Scientific Practice*. Cambridge, MA: MIT Press.

MacDougall, David. 1998. *Transcultural Cinema*. Princeton, NJ: Princeton University Press.

Martin, Emily. 1989. *The Woman in the Body: A Cultural Analysis of Reproduction*. Boston: Beacon Press.

Mol, Annemarie. 2002. *The Body Multiple: Ontology in Medical Practice*. Durham, NC: Duke University Press.

Mulvey, Laura. 1975. "Visual Pleasure and Narrative Cinema." *Screen* 16 (3): 6–18.

Outram, Dorinda. 1994. "Body Politics." *Oxford Art Journal* 17 (2): 130–131.

Petchesky, Rosalind Pollack. 1987. "The Power of Visual Culture in the Politics of Reproduction." *Feminist Studies* 13 (2): 263–292.

Pink, Sarah. 2009. *Doing Sensory Ethnography*. Thousand Oaks: SAGE.

Porter, Theodore M. 1996. *Trust in Numbers: The Pursuit of Objectivity in Science and Public Life*. Princeton, NJ: Princeton University Press.

Sudnow, David. 1978. *Ways of the Hand: The Organization of Improvised Conduct*. Cambridge, MA: Harvard University Press.

Treichler, Paula A., Lisa Cartwright and Constance Penley, eds. 1998. *The Visible Woman: Imaging Technologies, Gender, and Science*. New York: New York University Press.

Wade, David. 2008. "A Dissenting Voice as the Genome Is Sifted to Fight Disease." September 15, 2008. Accessed July 23, 2013. http://www.nytimes.com/2008/09/16/science/16prof.html?_r=1&ref=health&oref=slogin. Last accessed 4 November 2013.

Waldby, Catherine. 2002. "Biomedicine, Tissue Culture, and Intercorporeality." *Feminist Theory* 3 (3): 239–254.

Waldby, Catherine. 2006. *Tissue Economies: Blood, Organs and Cell Lines in Late Capitalism*. Durham, NC: Duke University Press.

Walsh, Maria. 2006. "Against Fetishism: The Moving Quiescence of Life 24 Frames a Second." *Film Philosophy* 10 (2): 1–10.

Contributors

Chiara Ambrosio is a lecturer in history and philosophy of science at the Department of Science and Technology Studies, University College London. Her research interests include visual culture and the relations between art and science in the nineteenth and twentieth century, American pragmatism and the philosophy of Charles S. Peirce, and general issues in history and philosophy of science.

Anne Beaulieu is the program manager of the Groningen Energy and Sustainability Program. Her ethnographically informed work explores the importance of interfaces for the creation and circulation of knowledge.

Andreas Birkbak is a PhD research fellow in the Techno-Anthropology Research Group at Aalborg University Copenhagen. His research is on the devices of publics. He is currently (spring 2014) a visiting scholar at École des Mines and Sciences Po, Paris.

Lisa Cartwright is a professor of communication and science studies at the University of California, San Diego, where she is also affiliated with critical gender studies. She works across visual studies; gender and sexuality studies; science, technology, information and medicine studies; and disability studies. Her most recent book is *Moral Spectatorship: Technologies of Voice and Affect in Postwar Representations of the Child* (Duke University Press, 2008). She is coauthor, with Marita Sturken, of *Practices of Looking: An Introduction to Visual Culture* (Oxford University Press, 2nd ed., 2008). Her earlier work on motion picture film and medicine is contained in *Screening the Body: Tracing Medicine's Visual Culture* (University of Minnesota Press, 1995). With Paula Treichler and Constance Penley, she coedited the volume *The Visible Woman: Imaging Technologies, Gender and Science* (NYU Press, 1998).

Annamaria Carusi is an associate professor in philosophy of medical science and technology studies at the University of Copenhagen. Her work aims to forge a new interdisciplinarity across humanities and social science approaches to the study of the natural and medical sciences. She has published extensively on imaging and visualization in science, and on computational modeling and simulation.

Matt Edgeworth is Honorary Research Fellow in archaeology and ancient history at the University of Leicester. He has written extensively on archaeological theory and practice. His books include *Acts of Discovery:* (2003), *Ethnographies of Archaeological Practice* (2007) and *Fluid Pasts: Archaeology of Flow* (2011).

Peter Galison is the Pellegrino University Professor in History of Science and Physics at Harvard University and director of the Collection of Historical Scientific Instruments. His main work explores the complex interaction between the three principal subcultures of twentieth-century physics—experimentation, instrumentation and theory. Galison is the author of several books in the history of science. Of particular relevance to this volume is the book coauthored with Lorraine Daston, *Objectivity* (Zone Books, 2007). His current work *Building Crashing Thinking* tracks the way technology restructures the self. Galison has cowritten and coproduced documentary films on the politics of science, including *Ultimate Weapon: The H-bomb Dilemma* (2000, with Pamela Hogan); *Secrecy* (2008, with Robb Moss); and, in preparation, *Containment* (2014), on the confinement of radiological materials.

Aud Sissel Hoel is an associate professor of visual communication in the Department of Art and Media Studies at the Norwegian University of Science and Technology, and the principal investigator on the interdisciplinary project *Picturing the Brain: Perspectives on Neuroimaging*. Her research explores the material conditions of mediation, focusing on photography, scientific imaging and technologies of thinking.

Torben Elgaard Jensen is a professor in techno-anthropology and STS, and the leader of the Techno-Anthropology Research Group at Aalborg University Copenhagen. His research focuses on digital methods in STS and user-oriented design practices.

Michael Lynch has written many books and articles over the past three decades on discourse, visual representation and practical action in research laboratories, clinical settings and legal tribunals. He was editor of *Social Studies of Science* from 2002 until 2012, and he was president of the Society for Social Studies of Science in 2007–2009. He is coeditor (with Catelijne Coopmans, Janet Vertesi and Steve Woolgar) of *Representation in Scientific Practice Revisited* (MIT Press, 2014).

Anders Koed Madsen is an assistant professor in the Techno-Anthropology Research Group at Aalborg University Copenhagen. His research focuses on digital methods and the use of Big Data as analytical tools in contemporary organizations.

Anders Kristian Munk is an assistant professor in techno-anthropology at Aalborg University Copenhagen and visiting researcher at the SciencesPo Medialab in Paris. His research focuses on knowledge controversies, risk issues, digital methods and visualization.

David Ribes is an assistant professor in the Communication, Culture and Technology program at Georgetown University. His research focuses on the development and sustainability of research infrastructure (i.e., networked information technologies for the support of science) and how it is transforming the practice and organization of contemporary knowledge production. His methods are ethnographic and archival. Please see http://davidribes.com

Kathryn de Ridder-Vignone is a postdoctoral research associate at the Center for Nanotechnology in Society at Arizona State University. In 2014 she will begin as an assistant professor of social context in the Department of Integrated Science and Technology at James Madison University. Her work examines how visual and material forms of knowing and learning shape nonscientists' scientific authority and the making of emerging socio-technical systems.

Tom Schilling is a doctoral candidate in the program in History, Anthropology, and Science, Technology, and Society at the Massachusetts Institute of Technology. Prior to his current research interests in geology, cartography and Canadian First Nations land claims negotiations, Schilling worked as a materials scientist, chemist and nanotechnologist. He has also written on the role of visualizations in materials science pedagogy and nanotechnology research and entrepreneurship, and on the implications of computer-assisted visualization in Brazilian science policy.

Alma Steingart is a junior fellow of the Harvard Society of Fellows. A historian of mathematics, she is currently completing two projects. The first investigates American pure and applied mathematics in the three decades following World War II; the second examines the efflorescence of new techniques by which mathematicians model abstract ideas in multiple media from 1970 to now.

Timothy Webmoor is an assistant professor adjunct in the Department of Anthropology, University of Colorado at Boulder, and a senior founding member of the Metamedia Lab, Stanford University. An ethnographer of science trained in archaeology, he both crafts new modes of documentation with digital media, and researches the implications of visualizing technologies for the political economy of academia. He's the co-author of *Archaeology: The Discipline of Things* (2012, University of California Press).

Steve Woolgar is the chair of Marketing and head of Science and Technology Studies at Saïd Business School, University of Oxford. With several books in science and technology studies (STS), the volume coedited with Michael Lynch in 1990, *Representation in Scientific Practice*, profoundly influenced this area of research. Woolgar's areas of expertise within the field of STS include governance and accountability relations, mundane objects and ordinary technologies, provocation and intervention, visualization and evidence, social theory and the use of neuroscience in business and management.

Albena Yaneva is a Professor of Architectural Theory and director of the Manchester Architecture Research Centre (MARC) at the University of Manchester, United Kingdom. She is the author of several books. Of particular relevance to this volume is the book *Mapping Controversies in Architecture* (2012, Ashgate). Yaneva is the recepient of the RIBA President's award for oustanding research (2010). Her research is intrinsically transdisciplinary and crosses the boundaries of science studies, cognitive anthropology, architectural theory and political philosophy.

Index